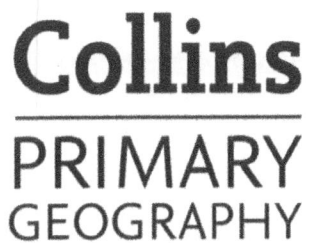

Issues

Teacher's Guide 6

Stephen Scoffham | Colin Bridge

Geography in the primary school	2
Collins Primary Geography overview	3
Places, themes and skills	4
Layout of the units	6
Lesson planning	8
Lesson summary	9
Studying the local area	10
Studying places in the UK and wider world	11
Differentiation and progression	12
Assessment	13
Ofsted and the National Curriculum in England	14
Support and guidance	15
Unit-by-unit notes	16
Photocopiable resource matrix	26
Photocopiable resources	30
Geography in the National Curriculum in England	60
World maps	62

Geography in the primary school

Geography is the study of the Earth's surface. It helps children understand the human and physical forces which shape the environment and the way it is changing. Children are naturally interested in their immediate surroundings. They also want to know about places beyond their direct experience. Geography is uniquely placed to satisfy this curiosity.

Geographical enquiries

Geography is an enquiry-led subject that explores fundamental questions such as:

- Where is this place?
- What is this place like (and why)?
- How and why is it changing?
- How does this place compare with other places?
- How and why are places connected?

These questions involve not only finding out about the natural processes which have shaped our environment; they also involve finding out how people have responded to them. Studying this interaction at a range of scales from the local to the global and asking questions about what is happening in the world around us lies at the heart of both academic and school geography.

Geographical perspectives

Geographical perspectives offer a uniquely powerful way of seeing the world. Since the time of the Ancient Greeks, geographers have been attempting to chronicle and interpret their surroundings. One way of seeing links and connections is to think in terms of key concepts and ideas. Three concepts, which have proved particularly useful in a range of settings, are place, space and scale.

- Place focuses attention on specific places (real and imagined) and highlights their character, current activities, changes and development.
- Space focuses attention on the relationship between features and places, and refers to where they are located, the patterns they form and networks connecting them.
- Scale enables geographers to look at the world from very small local sites to international regions.

A layer of additional concepts provides a further way of enhancing geographical understanding. These concepts include pattern, change, movement, interconnections, culture, power, sustainability and environmental impact. Taken in combination, these concepts act as a 'lens' for describing and analysing the complexities of the world around us.

As they conduct their enquiries and investigations geographers make use of some subject-specific skills. Foremost among these are mapwork and the ability to represent spatial information. Geographers also champion the use of digital data which enables them to portray changes and explore different scenarios. The use of maps, charts, diagrams, tables, sketches and other cartographic techniques – all of which allow us to visualise and better understand data – come under the more general heading of 'graphicacy'. Graphicacy is sometimes seen as a key human attribute and a distinguishing feature of geographical thinking.

Geography in primary schools gives children from the earliest ages a fascinating window onto the world. It embraces major concerns such as climate change, migration and biodiversity loss. The challenge for educators is to find ways of providing experiences and selecting content that will help children develop an increasingly deep understanding of the world around them.

Collins Primary Geography overview

Collins Primary Geography is a complete programme for pupils in the primary school and can be used as a structure for teaching geography from ages 5–11 and beyond. At its core are six Pupil Books, each of which has a linked Workbook. This Teacher's Guide provides teaching notes and photocopiable resources for each lesson. Editable Word, PDF and PowerPoint files are available to help you adapt the resources to the needs of your class. Audio files are also available for the stories in Pupil Books 1 and 2.

Aims

The overall aim of the programme is to inspire children with an enthusiasm for geography and to empower them as learners. The underlying principles include a commitment to international understanding in a more equitable world; a concern for the future welfare of the planet; and a recognition that creativity, hope and optimism play a fundamental role in lasting learning. Three different dimensions – connecting to the environment, connecting to each other and connecting to ourselves – are explored throughout the programme in different contexts and at a range of scales. We believe that learning to think geographically in the broadest meaning of the term will help to prepare children for the future and whatever it may hold.

Structure

Collins Primary Geography provides full coverage of the National Curriculum in England framework. Each Pupil Book covers a balanced range of themes and topics and includes case studies with a more precise focus:

- Book 1 *World around me* introduces children to the world at a local scale.
- Book 2 *Our planet* explores the world at a global scale.
- Book 3 *Investigation* encourages pupils to conduct their own research and enquiries.
- Book 4 *Movement* considers how movement affects the physical and human environment.
- Book 5 *Change* includes case studies on how places alter and develop.
- Book 6 *Issues* considers more complex ideas to do with the environment and sustainability.

Although the books are not limited to a specific age group, Book 1 will be particularly suitable for children at the beginning of their formal education. Book 2 is suitable for ages 6–7, Book 3 for ages 7–8, Book 4 for ages 8–9, Book 5 for ages 9–10 and Book 6 for ages 10–11, or children at the end of primary school.

The programme is structured in such a way that key themes are revisited, making it possible to investigate a specific topic in greater depth if required.

Investigations

Enquiries and investigations are an important part of pupils' work in primary geography. Asking questions and searching for answers can help children develop core knowledge, understanding and skills. Fieldwork is time-consuming when it involves travelling to distant locations, but local area work can be equally effective. Many of the exercises in *Collins Primary Geography* focus on the classroom, school building and local environment. We believe that such activities can have a seminal role in promoting long-term positive attitudes towards sustainability and the environment.

Places, themes and skills

Collins Primary Geography Books 3 to 6 follow a structure that gives a balance between places, themes and skills. In the opening units, pupils are introduced to topics including the physical geography of Planet Earth, water, weather, settlements, work and travel, and the environment. The units that follow take a global perspective and highlight a range of case studies focusing on the UK, Europe, North or South America, and Africa or Asia. Key geographical skills such as mapwork and fieldwork are featured in all the units. The overall aim is to provide a balanced coverage of geography.

Places

Locality studies are featured in each unit. These studies illustrate how people interact with their physical surroundings in a constantly changing world. They draw on first-hand accounts, focus on contemporary issues and highlight successful classroom activities. The places have been carefully selected to enable pupils to develop a framework of reference points which will enable them to place new knowledge in context by the time they have completed the scheme.

Themes

Physical geography is covered in the initial three units of each book which focus on Planet Earth, water and weather. Human geography is considered in units on settlements, and work and travel. There is also a unit specifically devoted to the urban and rural environment and human impact on the natural world. This is a very important aspect of modern geography and a key topic for schools generally. Climate change is considered in all the units to draw attention to its effects in many areas of our lives. Pupil Book 6 includes a dedicated unit exploring climate change and what we can do about it. When teaching children about climate-change issues, it is important that children are not left feeling helpless. Learning about current issues is the first step towards constructive engagement and action.

Skills

Maps and plans are introduced in context to convey information about the places being studied. The books contain maps at scales which range from the local to global. Charts, diagrams and other graphical devices are included throughout to illustrate a variety of techniques which children can emulate. Fieldwork is strongly emphasised, and all the books include projects, investigations and mapwork exercises which can be conducted in the local environment. Please note the usual fieldwork health and safety considerations before undertaking these activities.

Cross-curricular links

The different units in *Collins Primary Geography* can be easily linked with other subjects. The physical geography units have natural synergies with themes from sciences, as do the units on the environment. Local area studies overlap with work in history. Furthermore, the opportunities for promoting the core subjects are particularly strong. For example, the interpretation and presentation of data in various tables, graphs and models has clear links to work in mathematics. Each lesson is centred around discussion questions, and many of the investigations involve written work in different modes and registers.

Oracy and critical thinking

Each lesson provides an opportunity for pupils to engage with and explore information and ideas through discussion. Discussion panels present questions that are graded, with opening questions tending to be factual and later questions requiring critical thinking or personal input. These activities give all pupils the opportunity to practise speaking with confidence, explaining their ideas, listening and responding to others, and participating in group discussions. You, as the teacher, can facilitate whole-class, group or one-on-one discussions by modelling speaking and asking prompting questions. There are also opportunities for pair and groupwork in some of the mapwork and investigation exercises, as suggested in the unit-by-unit notes. These will further encourage development of oracy and critical thinking.

Places and themes for Books 3 to 6

Places and themes	Book 3 units	Book 4 units	Book 5 units	Book 6 units
Planet Earth	Landscapes	Coasts	Seas and oceans	Restless Earth
Water	Water around us	Rivers	Wearing away the land	Drinking water
Weather	Weather worldwide	Weather patterns	The seasons	Climate change
Settlements	Villages	Towns	Cities	Planning issues
Work and travel	Travel	Food and shops	Jobs	Transport
Environment	Caring for nature	Caring for towns	Pollution	Conservation
United Kingdom	Scotland	Northern Ireland	Wales	England
Europe	France	Germany	Greece	Europe
North and South America	South America	North America	North America	South America
Asia and Africa	Asia	Asia	Africa	Asia

Layout of the units

Books 3 to 6 each have ten units divided into three lessons. In earlier units, pupils are introduced to key themes based around Planet Earth, water, weather, settlements, work and travel, and the environment at increasing levels of complexity. Later units focus on places from around the UK, Europe, North or South America, and Africa or Asia. The overall aim is to provide a balanced coverage of geography.

Each lesson then follows a consistent layout, with several recurring features, as follows:

Unit title

Identifies the focus of the unit across the three lessons.

Lesson title

Identifies the theme of the lesson. The supporting Workbook unit and photocopiable resource use the same title which makes them easy to identify.

Enquiry question

Presents a focusing question for a whole section of the lesson, and suggests opportunities for open-ended investigations and practical activities.

Key words panel

Highlights key geographical words and terms which will be used during the lesson. Introduce these words as you teach the lesson. Use discussion to reinforce understanding. Children could build up geography notebooks over the course of a year (or longer). These might include key terms with supporting illustrations and/or definitions, if appropriate. These will be a valuable record of pupils' work and development.

Introductory text

Provides a simple introduction to the lesson's focus, presenting key knowledge opportunities. (The unit-by-unit notes provide further topic information to give you, as the teacher, additional context for teaching each lesson. These notes are not designed for the pupils).

Discussion panel

Consists of questions designed to draw pupils into the topic and to stimulate discussion. The first question often involves simple comprehension. Other questions involve reasoning and/or introduce a human element which helps to relate the topic to the child's own experience. Guidance on encouraging oracy and critical thinking through the facilitation of high-quality discussion is given in the unit-by-unit notes.

Photographs and graphics

Photos and illustrations provide opportunities for exploration, demonstration and discussion. Graphical devices ranging from maps to satellite images amplify the topic.

Data bank

Provides extra information to engage children and encourage them to find out more for themselves. Pupils can research additional facts and figures for themselves, use them in a quiz or game, or simply add them to their geography notebook or a class display.

Climate change panel

Provides extra information and further discussion opportunities to show children the effects of climate change in many areas of our lives.

Mapwork exercise

Indicates how the lesson can be developed through atlas and mapwork.

Investigation panel

Suggests a practical activity which will help pupils consolidate their understanding.

Summary panel

Indicates the knowledge and understanding covered in the unit.

Photocopiable resources

Each lesson has a supporting photocopiable resource. These explore the concepts and ideas which underpin the unit, and extend what is presented in the book in engaging and informative ways. The resources may also be used to consolidate and assess understanding of the key concepts. See pages 30–59 of this book.

Workbook

The Workbook includes additional activities for each lesson, supporting learning, providing scaffolding for investigations, encouraging pupils to apply their learning and allowing teachers to provide evidence of learning in geography for each child.

Layout of the units

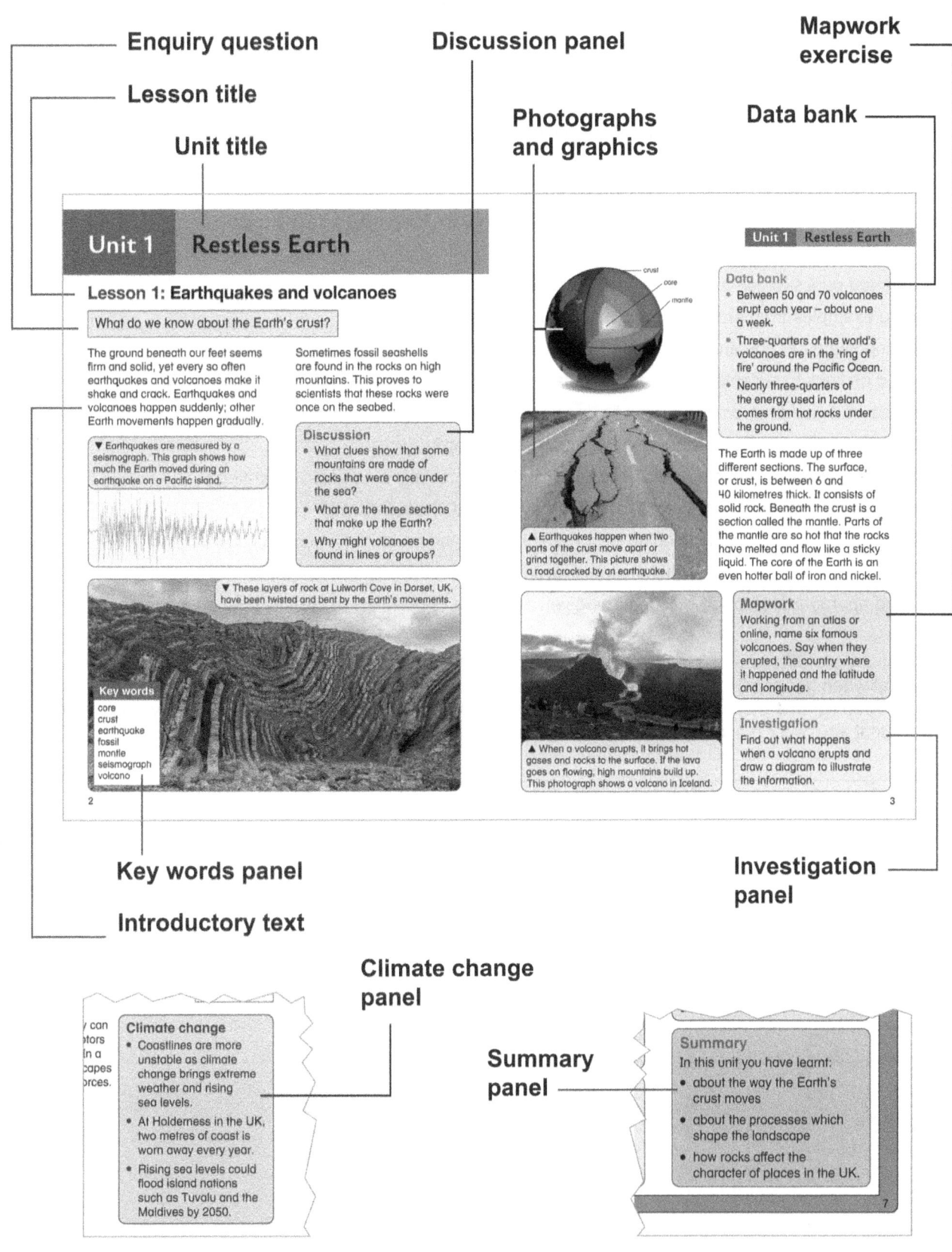

Lesson planning

Collins Primary Geography has been designed to support both whole-school and individual lesson planning. As you devise your schemes and work out lesson plans, you may find it helpful to ask the following questions. Have you:

- Given children a range of entry points which will engage their enthusiasm and capture their imagination?
- Used a range of teaching strategies which cater for pupils who learn in different ways?
- Thought about how you will draw pupils into class discussion, supporting them to develop their understanding and ideas?
- Thought about using practical activities and games?
- Explored the ways that stories or personal accounts might be integrated with the topic?
- Considered the opportunities for fieldwork?
- Encouraged pupils to use and make maps and diagrams?
- Included examples from around the world to enhance global awareness?
- Questioned whether you are challenging rather than reinforcing stereotypes?
- Checked on links to suitable websites, particularly with respect to research?
- Made use of digital programs to record findings or analyse information?
- Made links to other subjects where there is a natural overlap?
- Promoted geography alongside oracy and literacy skills especially in talking and writing?
- Taken advantage of the opportunities for presentations and class displays?
- Ensured that the pupils are developing geographical skills and meaningful subject knowledge?
- Clarified the knowledge, skills and concepts that will underpin the lesson?
- Identified appropriate learning outcomes or given pupils the opportunity to identify their own ones?

These questions are offered as prompts which may help you to generate stimulating and lively lessons. There is clear evidence that when geography is fun and pupils enjoy what they are doing, it can lead to lasting learning. Striking a balance between light-hearted delivery and serious intent is part of the craft of being a teacher. *Always remember to follow the latest advice for practising online safety in research activities.*

Finding time for geography

The pressures on the school timetable and the demands of the core subjects make it hard to secure adequate time for primary geography. However, finding ways of integrating geography with mathematics and literacy can be a creative way of increasing opportunities. Geography also has a natural place in a wide range of social studies and current affairs whether local or global. It can be developed through class assemblies and extra-curricular studies. Those who are committed to thinking geographically find a surprising number of ways of developing the subject whatever the accountability regime in which they operate.

Lesson summary

The table below provides an overview of the lessons in *Collins Primary Geography Pupil Book 6*. Individual schools may want to adapt the lessons and associated activities according to their particular needs and circumstances.

Theme	Unit focus	Lesson 1	Lesson 2	Lesson 3
Planet Earth	Restless Earth	Earthquakes and volcanoes	Creating landscapes	Rocks and soils in the UK
Water	Drinking water	Water, water everywhere	Water supplies	Conserving water
Weather	Climate change	Global warming	Unusual weather	Responding to climate change
Settlements	Planning issues	Reasons for development	Old sites, new uses	Planning game
Work and Travel	Transport	Travelling further, travelling faster	Transport problems	Hidden costs
Environment	Conservation	Threatened wildlife	Antarctica	Conservation projects
United Kingdom	England	Introducing England	Finding out about Sandwich	Living in Sandwich
Europe	Europe	Introducing Europe	The European Union	Celebrating Europe
North and South America	South America	Introducing the Amazon	Using the rainforest	Saving the Amazon
Asia and Africa	Asia	Southeast Asia	Investigating Singapore	A Singapore family

Studying the local area

The local area is the immediate vicinity around the school and the home. It consists of three different components: the school building, the school grounds, and local streets and buildings. By studying their local area, children will learn about the different features which make their environment distinctive and how it attains a specific character. When they are familiar with their own area, they will then be able to make meaningful comparisons with more distant places.

There are many opportunities to support the lessons outlined in *Collins Primary Geography* with practical local area work. First-hand experience is fundamental to good practice in geography teaching, is a clear requirement in the programme of study and has been highlighted in guidance to Ofsted inspectors. The local area can be used not only to develop ideas from human geography but also to illustrate physical and environmental themes. The checklist below illustrates some of the features which could be identified and studied.

Physical geography

Hill, valley, cliff, mountain, rock, slope, soil, forest

River, stream, pond, lake, estuary, ocean, sea, beach, coast

Slopes, rock, soil, vegetation and other small-scale features

Local weather, seasons and site conditions

Human geography

Origins of settlements (city, town, village), land use (farms) and economic activity

House, cottage, terrace, flat, housing estate

Roads, stations, harbours, ports

Shops, factories and offices

Fire, police, ambulance, health services

Library, museum, park, leisure centre

All work in the local area involves collecting and analysing information. An important way in which this can be achieved is through the use of maps and plans. Other techniques include annotated drawings, bar charts, tables and reports. There will also be opportunities for the children to make presentations in class and perhaps to the rest of the school in assemblies.

Misconceptions

There is a growing body of research which helps practitioners to understand more about how children learn primary geography and the barriers and challenges that they commonly encounter. The way that young children assume that the physical environment was created by people was first highlighted by Jean Piaget. The importance and significance of early childhood misconceptions was further illuminated by Howard Gardner. More recent research has considered how children develop their understanding of maps and places. Children's ideas about other countries and their attitudes to other nationalities form another very important line of enquiry. So, too, do their ideas about climate change and the environment. Some key readings are listed in the references on page 15.

Studying places in the UK and wider world

Collins Primary Geography Pupil Book 6 contains detailed case studies of the following places in the UK and around the world. Place studies focus on small-scale environments and everyday life, which means they relate to children's needs and understanding.

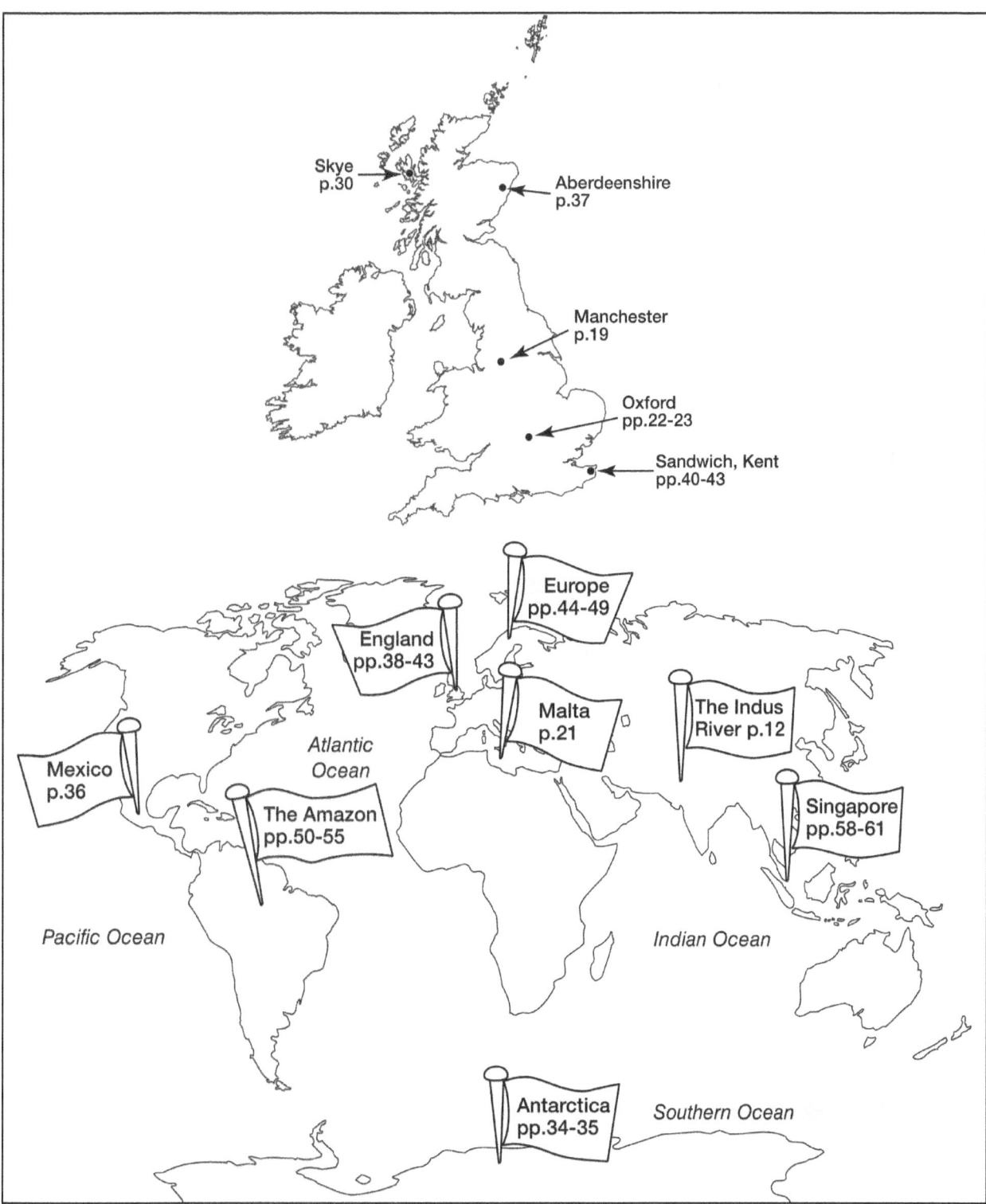

Further examples from around the world are also included throughout the units. By considering people and describing their surroundings, the information is presented at a scale and in a manner which relates particularly well to children. Research shows that pupils tend to reach a peak of friendliness towards other countries and nations at about the age of 10. It is important to capitalise on this educationally and to challenge prejudices and stereotypes.

Differentiation and progression

Collins Primary Geography sets out to provide access to the curriculum for children of all abilities. It is structured so that children can respond to and use the material in a variety of ways.

- Each lesson contains a range of stimulus material designed to engage children imaginatively. This means activities can be selected which are appropriate to individual circumstances.
- Considerable emphasis is placed on discussion and critical thinking appropriate for the age group, which allows teachers to frame discussions and respond to pupils according to their level of understanding.
- The Workbooks provide further opportunities to develop and evidence understanding.
- The photocopiable resources allow you to assess understanding of key concepts during the unit.

Teachers will be able to select what they think will be appropriate from a range of resources. There is no need to work through all the material.

Differentiation by outcome

Each lesson starts with an introductory text and linked discussion questions designed to capture the children's imagination and draw them into the topic. There are opportunities for children who require more support to relate the material to their own experience. Other children will be able to consider the underlying geographical concepts. The pace and range of the discussion can be controlled to suit the needs of the class or group.

Differentiation by task

The mapwork and investigation exercises can be modified according to the pupils' ability levels. Teachers may decide to complete some of the tasks as class exercises or help learners who require more support by working through the first part of an exercise with them. Classroom assistants could also work with individual children or small groups. Some children could be given extension tasks. Ideas and suggestions for extending each lesson are provided in the information on individual lessons (pages 16–25). The Workbooks include a wider range of activities with varying levels of scaffolding to support Pupil Book investigations when required.

Differentiation by process

Children of all abilities benefit from exploring their environment and conducting their own investigations. The investigation activities include many suggestions for direct experience and first-hand learning. Work in the local area can overcome the problems of written communication by focusing on concrete events. There are also opportunities for taking photographs and conducting surveys as well as for making lists, diagrams and written descriptions.

Progression

The themes, language and complexity of the material have been graded to provide progression between each title. However, the gradient between different books is deliberately shallow. This makes it possible for the books to be used interchangeably by different year groups or within mixed-ability classes. The way that this might work can be illustrated by considering a sample unit. For instance, in Book 3 the unit on weather introduces children to hot and cold places around the world. Book 4 looks at ways of recording the weather; Book 5 focuses on the seasons, and Book 6 considers the effects of climate change. This approach provides opportunities for reinforcement and revisiting which will be particularly helpful for children who require more support.

Assessment

Assessment is often seen as having two different dimensions.

- *Formative assessment* is an ongoing process which provides both pupils and teachers with information about the progress they are making in a piece of work.
- *Summative assessment* occurs at defined points in a child's learning and seeks to establish what they have learnt and how they are performing in relation both to their peers and to nationally agreed standards.

Collins Primary Geography provides opportunities for both formative and summative assessment.

Formative assessment

- The discussion questions invite pupils to discuss a topic, relate it to their previous experience and consider any issues which may arise, thereby yielding information about their current knowledge and understanding.
- The mapwork exercises focus especially on developing spatial awareness and skills and will indicate the pupils' current level of ability.
- The investigation activities give pupils the chance to extend their knowledge in ways that match their current abilities.
- The Workbook activities provide an opportunity for pupils to apply their understanding and evidence learning.

Summative assessment

- The Summary panels at the end of each unit in the Pupil Book highlight key learning outcomes. These can be tested directly through individually designed exercises. These unit aims are repeated in the unit-by-unit notes in this Teacher's Guide.
- The photocopiable resources (see pages 30–59) can be used to provide further evidence of key concepts, and to track progress of knowledge and skills. Whether used formatively or summatively, they are intended to consolidate understanding and identify gaps in learning to inform future teaching.

Reporting to parents and guardians

Collins Primary Geography is structured around geographical skills, themes and places. As children work through the lessons, they can build up a folder of work and progress through the Workbook. This will provide evidence of mapwork and other practical activities both inside and outside the classroom and provide a rounded portrait of pupil achievement. This will also be a useful resource when teachers report to parents and guardians and show if a child requires further support in geography.

National curriculum reporting

There is a single attainment target for geography and other National Curriculum subjects in the National Curriculum for England. This simply states that

'By the end of each key stage, pupils are expected to know, apply and understand the matters, skills and processes specified in the relevant programme of study.'

This means that assessment need not be an onerous burden, and that evidence of pupils' achievement can be built up over a set of years (or a Key Stage). The assessment process can also inform lesson planning.

Ofsted and the National Curriculum in England

The National Curriculum in England provides a framework for geography but doesn't specify the details of what should be taught or in what depth. Schools have the flexibility to choose their own curriculum approaches provided they pay sufficient attention to (a) context (b) structure (c) sequencing and (d) implementation. There is a significant emphasis on factual knowledge. Ofsted argue that it is essential to identify what children need to remember and to use it transferably in different circumstances.

The curriculum

The curriculum refers to what is taught and should support children to build their knowledge over time. Ofsted distinguish between different forms of knowledge:

- *Substantive knowledge* refers to knowledge relating to the themes and topics specified in the National Curriculum. This could be seen as the 'vocabulary' of geography.
- *Disciplinary knowledge* involves applying a geographical 'lens' to a particular or area of study and can be regarded as the 'grammar' of geography.

If pupils haven't grasped substantive geographical knowledge, then they will be unable to think or speak geographically.

Lesson planning

Ofsted argues that pupils get better in geography by building on their prior knowledge and applying it in new more complex ways (Ofsted 2021). Activities should therefore be selected which help pupils to build their knowledge and consolidate what they have already learnt in time-efficient ways. This draws attention to the importance of sequencing in geography. It is important to think carefully not only about the building blocks of geography but also about what comes after as well as what comes before any particular topic. In this way the curriculum becomes the assessment model.

Inspection findings

Inspection findings indicate that practice is not always as good as it could be. Areas of weakness include mapwork, fieldwork, sequencing and the application of geographical concepts (Ofsted 2021, 2023). To some extent this is unsurprising given that limited time has been spent learning how to teach geography during primary training (Ofsted 2023). However, there is also clear evidence that pupils enjoy geography and are curious about the world around them. The fact many are passionate about the Earth and the need to care for it also attracted very favourable comment from inspectors (Freeland 2021).

These prompts may help you prepare for an inspection:

- Identify a teacher who is responsible for developing the geography curriculum.
- Decide how geography will fit into your whole school plan.
- Make an audit of current geography teaching to identify gaps and weaknesses.
- Discuss and develop a geography policy which includes statements on overall aims, topic planning, teaching methods, progression, assessment and recording.
- See that all members of staff are familiar with the geography curriculum.
- Organise in-service training to rectify any areas of weakness.
- Review and update geography teaching resources.
- Devise an action plan for geography which includes an annual review procedure.
- Discuss the policy with the school governors.
- Provide a regular opportunity for discussing geography teaching in staff meetings.

Support and guidance

Primary Geography Quality Mark

The Primary Geography Quality Mark set up by the UK Geographical Association is a self-assessment framework designed to help subject leaders. There are three categories of award. The 'bronze' level recognises that lively and enjoyable geography is happening in your school; the 'silver' level recognises excellence across the school; and the 'gold' level recognises that excellence that is shared and embedded in the community beyond the school. The framework is divided into four separate cells: (a) pupil progress and achievement; (b) quality of teaching; (c) behaviour and relationships; (d) leadership and management. For further details see the Geographical Association website.

Achieving accreditation for geography in school is a useful way of badging achievements and identifying targets for future improvement. This makes it an effective and efficient way of raising standards. The Geographical Association provides a wide range of support to teachers to help with this process. In addition to conferences and CPD sessions it produces a journal for primary schools, *Primary Geography*, three times a year. See the Geographical Association website for full details of the books and guides it publishes for classroom use.

Networking, training and sharing ideas

Networking and sharing ideas can happen on an informal basis amongst friends and professional colleagues. Conferences and CPD training and events provide a more formal way of developing and extending your knowledge of geography teaching. The support that comes from networking will help you to grow in confidence and broaden your ideas. Speaking or writing about what you have been doing will further consolidate your ideas. Subject associations, environmental organisations and development education centres usually welcome new members with enthusiasm (for example, see the Geographical Association and the Royal Geographical Society websites). The sense of community that they foster cannot be underestimated.

References and reading

Barlow, Anthony, and Sarah Whitehouse (2019) *Mastering Primary Geography*, London: Bloomsbury Academic.

Bonnett, Alistair (2023) *What is Geography?* (2nd edn), Lanham, Maryland: Rowman and Littlefield.

Cannell, Jon (2023) 'Geographical concepts in primary education', *Primary Geography*, 112: pp8–9.

Catling, S. et al. (2022) 'Aspiring to High-Quality Primary Geography: A report on a study of the GA's Primary Geography Quality Mark Moderators' feedback to schools', Sheffield: Geographical Association.

Dolan, Anne M. (2020) *Powerful Primary Geography: A Toolkit for 21st-Century Learning*, London: Routledge.

Freeland, Iain (2021) 'Geography in outstanding primary schools', Ofsted: schools and further education & skills (FES) blog, gov.uk.

Ofsted (2021) 'Research review series: Geography', gov.uk.

Ofsted (2023) 'Getting our bearings: geography subject report', gov.uk.

Roberts, Margaret (2023) *Geography Through Enquiry: Approaches to teaching and learning in the secondary school* (2nd edn), Sheffield: Geographical Association (Chapter 4).

Scoffham, Stephen (ed.) (2016) *Teaching Geography Creatively (Learning to Teach in the Primary School Series)* (2nd edn), London: Routledge.

Scoffham, Stephen and Paula Owens (2024) *Bloomsbury Curriculum Basics: Teaching Primary Geography* (2nd edn), London: Bloomsbury.

Tanner, Julia and Stephen Pickering (eds) (2017) 'Taking the Learning Outdoors at KS1', *Teaching Outdoors Creatively*, London: Routledge.

Trait, Georgie, et al. (2024) 'The key ingredients for quality geography', *Primary Geography*, 114: p19.

Willy, Tessa (ed.) (2019) *Leading Primary Geography: The essential handbook for all teachers*, Sheffield: Geographical Association.

Unit-by-unit notes

Unit 1: Restless Earth

> **In this unit, pupils learn:**
> - about the way the Earth's crust moves
> - about the processes which shape the landscape
> - how rocks affect the character of places in the UK.

Earth tremors can happen almost anywhere in the world. However, most tremors and earth movements occur within well-defined earthquake belts as the different plates which make up the Earth's crust move around. The vibrations are recorded on a machine called a seismograph, and the amount of energy released is described using the Richter scale.

Lesson 1: EARTHQUAKES AND VOLCANOES

What do we know about the Earth's crust?
This lesson introduces the idea that the Earth's crust, whilst it may appear solid, is actually constantly moving, albeit very slowly. Earth tremors, whether from earthquakes or volcanoes, are common occurrences, although most are very minor.

Discussion
Pupils should be able to answer the first two questions fairly easily. The final question is more challenging. Volcanoes and earthquakes often occur along the fault lines where different parts of the Earth's crust move apart or under each other.

Mapwork
Finding out the names and locations of a few famous volcanoes may lead pupils to want to discover more. Always follow the latest advice for practising online safety in research activities.

Investigation
There are many different types of volcano with different shapes and structure. The classic example is a cone made from layers of ash and lava.

Lesson 2: CREATING LANDSCAPES

What forces shape the land?
Rocks are worn away very slowly over long periods of time. Even adults find it difficult to understand how small changes can eventually reduce mountain ranges to sea level or can build up the coastline. It is enough at this stage to simply introduce pupils to the general idea.

Discussion and Climate change
In landscape terms, the impact of climate change is particularly evident along the coast. Higher temperatures are causing seawater to expand and polar ice to melt bringing the risk of flooding to low-lying areas. Climate change is also resulting in more severe storms and powerful waves, which are reshaping beaches and undermining cliffs. You can support children's discussion with this information.

Investigation
The survey of wear and tear in the school grounds is a way of introducing pupils to the idea of erosion.

Lesson 3: ROCKS AND SOILS IN THE UK

How has the landscape of the UK formed?
There are plenty of clues which reveal the geological history of the UK. In the mountains of Wales, Scotland and the Lake District, for example, glaciers have shaped the U-shaped valleys and sharp ridges. Elsewhere, fossils of sea creatures indicate that rocks now on dry land were once under water. Unusually, the UK has rocks dating from nearly all periods of geological history. This lesson presents a simplified portrait of key periods.

Rocks in the street
Thinking about how rocks are used as building materials provides natural links to the study of materials and their properties in science.

Mapwork
Many of us don't ever think about the rocks that are under our feet, in our buildings or in the goods we buy. Pupils will not need to go too far when they devise their trails.

Investigation
Children often enjoy contributing items they have collected to a display table. In addition to a variety of colours and textures, fossils give a rock collection added interest and provide a good opportunity for discussion.

Climate change
Rocks provide tangible evidence of Earth's climate at different times in the past. Using fossils and radioactive dating techniques, scientists are able to establish their age. Their structure and composition provide valuable evidence of the conditions in which they were formed.

> **Teacher's Guide photocopiable resources**
> Use pages 30–32 to consolidate key concepts.
>
> **Workbook**
> See pages 2–7 for additional supportive activities.

Unit-by-unit notes

Unit 2: Drinking water

In this unit, pupils learn:
- where drinking water comes from
- that polluted water causes illness
- how people can save water.

Water, like food, is a basic human need. Each person needs a minimum of about 50 litres of water a day to sustain a reasonable quality of life. Water consumption varies enormously between countries and different communities. Around the world only about five per cent of water is used for domestic purposes. Industry takes 20 per cent, and the rest is used for irrigation. The demand for water is rising as living standards rise and population increases. Climate change is also making some areas drier, adding to water-supply problems.

Lesson 1: WATER, WATER EVERYWHERE

Is there enough water in the world?
In recent years there has been increasing interest in water security as people have come to recognise that fresh water is a finite resource. Part of the problem is that supplies are irregular. Periods of drought and water scarcity can alternate with floods and heavy rain. Finding ways to store surpluses is one way of increasing protection against variations in weather and climate.

Mapwork
You might like to talk through a map of your area with mountains, rivers and reservoirs feeding into the water supply before pupils work on this activity.

Investigation
Pupils may find it hard to visualise the data presented in the tap diagram. It will help them to relate to the figures if you fill a litre bottle with water as a demonstration.

Lesson 2: WATER SUPPLIES

Why is clean water so important?
Polluted water carries disease. In the United Kingdom, it was responsible for serious outbreaks of cholera in the 19th century. Dirty water and poor sanitation continues to be a severe problem in many places around the world today.

Discussion
Pupils might work in groups to discuss what they might do with these limited water supplies. This might be something to debate and justify within the group.

Mapwork
The diagrams of water supply will show the route the water takes and therefore are a form of map.

Investigation
This exercise touches on the sensitive issue of disparities around the world. As pupils learn about global inequalities, it is important to avoid promoting negative stereotypes or generalisations.

Lesson 3: CONSERVING WATER

Are we using water wisely?
The main idea in this lesson is that water needs to be conserved and used wisely. The case study sets the scene by focusing on the River Indus. As long as 5000 years ago, the city of Harappa had its own piped water supply, as the photograph of the storage tank illustrates. Today, water from the Indus irrigates large areas of Pakistan and supports a large proportion of its agriculture.

Discussion
You might check that pupils understand the scale of the River Indus and the vast lands it drains by looking at a regional map. This will lead into the mapwork activity.

Mapwork
Pupils might look online to see what information different maps of the Indus provide before making their sketch map.

A water survey
The survey illustrates how water can be conserved. You could develop the work by asking the children to write a report on ways of saving water in your school.

Investigation
The idea of making 'wise decisions' is an important one for environmentalists. It acknowledges that people have a legitimate claim to water resources but encourages restraint.

Teacher's Guide photocopiable resources
Use pages 33–35 to consolidate key concepts.

Workbook
See pages 8–13 for additional supportive activities.

Unit-by-unit notes

Unit 3: Climate change

In this unit, pupils learn:
- how the world is getting warmer
- how global warming is impacting people, plants and creatures
- some of the things you do in response to climate change.

Global warming was first identified by Joseph Fourier in the 1820s. He calculated that the atmosphere was keeping the Earth surprisingly warm. We now know that global warming is caused by a mixture of carbon dioxide, nitrous oxide, methane, and other gases which trap heat from the sun and stop it radiating back into space. Teaching children about climate change needs to be handled sensitively to avoid triggering eco-anxiety and develop fears of the unknown. This unit counters this with examples of positive action that we can take in response.

Lesson 1: GLOBAL WARMING

How is the world's climate changing?
Temperature records show a steady rise in global temperatures, especially in recent decades. An increase of one or two degrees centigrade may not sound very much, but the Earth is now warmer than it has been for thousands of years.

The greenhouse effect

Climate change
Discuss the diagram and information with children. Ensure they understand how each point in the diagram relates to rising temperatures. Talk about positive action that we can take in response to this. Lesson 3 focuses in more detail on how we can respond to climate change.

Mapwork
Generally, land areas tend to heat and cool more quickly than seas and oceans, which have moderate temperatures. What is happening in your part of the world? Discuss this as a class.

Discussion
The collapse of the Greenland ice sheet would cause sea levels to rise, flood low-lying areas and make people homeless. Burning fossil fuels provides cheap energy for homes and factories and fuel for cars.

Lesson 2: UNUSUAL WEATHER

What are the impacts of global warming?

Climate change
The physical, human and ecological impact of climate change are highlighted in these three bullet points. Can pupils think of one other example of each?

Mapwork
Small island states are particularly at risk from global warming and some, like Tuvalu, could become completely engulfed by water in the next 50 to 100 years. These island nations have combined to bring their plight to the attention of the international community.

Investigation
The decline in wildlife and the collapse of biodiversity are arguably every bit as serious as climate change. It is now generally agreed that we have entered a period of mass extinctions. This will be the sixth mass extinction in the history of our planet and the first caused by human activity. Pupils could investigate conservation efforts as well as threats. Always follow the latest advice for practising online safety in research activities.

Discussion
Pupils could work in small groups with a chair, who makes sure all contribute, and a note-taker, who will feed ideas back to the class.

Lesson 3: RESPONDING TO CLIMATE CHANGE

What can we do about climate change?

There is no single action we can take to combat climate change. Rather it requires a range of responses. Take time to explore the net zero graphic at the bottom of Pupil Book page 18. As well as deciding what each symbol represents, you might want pupils to come up with their own designs.

Investigation
This requires children to look back at the present from the perspective of the future, and to consider how future generations might judge our actions.

Discussion
Talk about the idea of a carbon footprint before pupils start to consider how they can reduce their personal impact. Make it clear too that they don't have to do everything at once or feel unnecessary guilty about their lifestyle. The wider point is that even small actions make a difference. Climate change is not something we can solve individually, but doing nothing is corrosive whilst taking action is empowering.

Teacher's Guide photocopiable resources
Use pages 36–38 to consolidate key concepts.

Workbook
See pages 14–19 for additional supportive activities.

Unit-by-unit notes

Unit 4: Planning issues

In this unit, pupils learn:
- that people want to use land in different ways
- how planning decisions are made
- how to obtain information from maps and aerial photographs.

As the pressure on land has increased in the UK, it has become necessary to introduce more and more planning laws. Until the early years of last century, people could build houses and factories almost anywhere they wanted. The first legislation was mainly concerned with controlling disease. Gradually, local authorities have been given powers to control how land is used. Since 1947, all new development needs to have planning permission. National parks, and urban and rural conservation areas have also been designated. Children are introduced to urban planning issues in this unit.

Lesson 1: REASONS FOR DEVELOPMENT

Why are there conflicts over land use?
The illustration on Pupil Book page 20 is a visual representation of some of the different ways in which people compete to use the same space. Reconciling different interests is a complicated process. Once land is used for development, it rarely returns to a natural state.

Investigation
Looking at old maps can be a fascinating activity – much will depend on the documents you can find.

Living on an island
With a population of over 1000 people per square kilometre, Malta is one of the most densely populated countries in Europe. The case study emphasises the idea that land is a finite resource and that people have to live within their means.

Discussion
These are thinking and discussing/debating questions that could be done in small groups. Ask pupils to consider a range of options and give reasons for their ideas.

Climate change
Planning policy is now taking more and more account of climate change. Tourism and farming are activities that are particularly vulnerable to extreme weather. You might want to discuss with your pupils what measures would make your own area more resilient to extreme weather.

Lesson 2: OLD SITES, NEW USES

How can old sites be redeveloped?
The redevelopment of the car factory site at Cowley in the UK illustrates how needs change over time and how public consultation is part of the planning process. It also shows how new developments do not always have to be put on greenfield sites. Redeveloping so-called 'brownfield' sites is one way of conserving land.

Discussion
This discussion will provide good preparation for the writing activity that follows in the Investigation. You might want to put pupils into groups so they can debate the advantages and disadvantages of different plans. You could extend the activity with a similar debate about what pupils would like to see in a new school building.

Investigation
The key point in this activity is that there were a number of different options at Cowley and that choices had to made between them.

Lesson 3: PLANNING GAME

How are planning decisions made?

Mapwork and Discussion
As well as developing pupils' map-reading skills, the Discussion and Mapwork exercises on Pupil Book page 24 highlight the different features and facilities in an urban area.

Investigation and Mapwork
Simulations and role plays can be a particularly effective way of engaging children and are a feature of good practice in primary geography education. The planning game in the Mapwork on page 25 might be developed further by involving outsider advisers such as an architect, builder or planner to act as consultant to advise pupils on their proposals.

To facilitate the Investigation, encourage pupils to think 'geographically' as they make their advertisements by using appropriate language and including maps and plans.

Teacher's Guide photocopiable resources
Use pages 39–41 to consolidate key concepts.

Workbook
See pages 20–25 for additional supportive activities.

Unit-by-unit notes

Unit 5: Transport

In this unit, pupils learn:
- that traffic problems are difficult to solve
- about different schemes to control traffic
- how people can change their travel habits.

The speed and availability of travel has grown phenomenally over the last century. In the 1920s, for example, it took almost a week to travel from the UK to the US by boat. Now this same journey can be completed in a matter of hours. Mass air travel has allowed people to move round the world as never before. Holiday resorts in places as far apart as Thailand and Mexico attract visitors from Europe and North America. At the same time, the growth of trade and international finance has linked different economies together in a single system. Globalisation has changed the way that people live their lives and made them increasingly interdependent.

Lesson 1: TRAVELLING FURTHER, TRAVELLING FASTER

What are the opportunities for travel in the world today?
This first lesson focuses on Europe to introduce pupils to international and global transport networks, and considers the pros and cons of mass travel.

Investigation
Once they have made up a word search for European cities, pupils could make up searches on other geographical themes as a homework or extension activity.

Discussion
Accessibility, price, distance, speed and environmental impacts are considerations to draw out.

Mapwork
As well as researching an air route map, you could ask the children to say what it shows. Are some places better connected than others? Which other continents have the most air routes? Is there more than one way to get to the same place?

Lesson 2: TRANSPORT PROBLEMS

Can roads cope with more traffic?
Coping with traffic has been a major challenge for many decades. Building motorways and bypasses was once seen as a way of solving traffic problems, but it is now recognised that new roads also attract traffic and create problems in other places. Similar difficulties affect air travel as many air routes are now operating at nearly full capacity.

The flow diagram on Pupil Book page 29 highlights how traffic management involves a combination of strategies rather than individual initiatives that provide one-stop solutions.

Discussion
This might work as a group exercise with pupils taking roles in a government transport meeting. Refer pupils to the traffic-management solutions used over the years in the flowchart on page 29. Have a chair and a note-keeper, followed by a report-back session to the class.

Investigation
The photograph of the lorry highlights the tension between our desire for cheap goods and mass transport and the impact of vehicles on people and the environment.

Mapwork
In addition to recording physical objects such as bollards and fences, pupils may want to annotate their plans to record parking regulations and other rules.

Lesson 3: HIDDEN COSTS

How do vehicles affect people and the environment?

You might give this lesson a positive slant by getting pupils to think of solutions rather than dwelling on the problems that vehicles cause.

Discussion
Pupils might consider manufacturing resources and emissions as well as fuel and maintenance. The Skye bridge concerns might include environmental damage, loss of wildlife, fumes from traffic, overtourism. Ask pupils to weigh up and justify their choice of worst problem.

Finding out about local transport

Investigation
You may need to adapt the questionnaire to match your local circumstances. Extend the work by asking the children to make a map of safe routes to school. They could also write a short report about any changes they would like to see.

Teacher's Guide photocopiable resources
Use pages 42–44 to consolidate key concepts.

Workbook
See pages 26–31 for additional supportive activities.

Unit-by-unit notes

Unit 6: Conservation

In this unit, pupils learn:
- why wildlife is threatened
- how people and countries can co-operate to protect the environment
- how farmers can use the land without harming it

The battle to save endangered species has a long history. Decades ago, environmental groups launched campaigns to save individual animals, such as whales, tigers and rhinos. Although these campaigns achieved considerable success, it became clear that much more was needed to safeguard the survival of other, less glamorous, life forms. For this reason attention switched to preserving whole environments, such as rainforests, ancient woodlands and coral reefs. Whole habitats and ecosystems have the advantage of containing a balance of life within them. The problem is that people will often only choose to preserve these environments if they also bring social and economic benefits. This has led to the notion of sustainable development, whereby people try to use the environment without upsetting its ecological balance. Making wise choices which enable us to live within the carry capacity of the planet is now seen as a key challenge for the 21st century. The stakes are getting steadily higher – over 70% of global wildlife has been lost over the last 50 years.

Lesson 1: THREATENED WILDLIFE

Why are many plants and animals endangered?

Discussion
Linked ecosystems and habitats will be central to these discussions.

Investigation
Many conservation groups have areas on their websites for children. Always follow the latest advice for practising online safety in research activities.

Mapwork
Get the children to find out about endangered creatures in different parts of the world. It is important to establish that conservation is a global challenge and doesn't just apply to the rainforest or distant places.

Lesson 2: ANTARCTICA

Why should Antarctica be conserved?
Antarctica is a unique continent which has been protected from development because of its harsh climate and isolated position. Its future as a pristine environment depends on international co-operation. The mixture of ecological, social and political issues which surround its protection make for a particularly interesting case study.

Discussion
Prompt pupils to think about the mixture of ecological, social and political issues that surround Antarctica. You might question them as they talk to bring these out. The arguments for and against a world park are considered in the second part of the lesson.

Investigation
Although Amundsen's expedition was the first to reach the South Pole, Captain Scott's story is more well-known in the UK.

Lesson 3: CONSERVATION PROJECTS

What are people doing to conserve the environment?
This lesson starts by focusing on the Monarch butterfly, which migrates from the United States and Canada to the forests of Mexico each winter.

Discussion
Butterfly numbers were declining due to deforestation and climate change. Involving local people and giving them jobs – by encouraging eco-tourism – has successfully stopped the decline in butterfly numbers. Instead of chopping down the forests, people see them as an asset. With money now coming in through eco-tourism, conservation project-work is thriving.

How can we keep a balanced environment?
The second part of the lesson provides a simple portrait of an organic farm. Both case studies show how people are supporting wildlife and finding ways of working with, rather than against, nature.

Mapwork
There are sites within walking distance of most schools which have the potential to be improved as habitats for plants and creatures. The opportunities for links with science makes this a good fieldwork activity.

Investigation
Exploring the advantages and disadvantages of organic farming will give children the opportunity to begin to formulate their own views.

Teacher's Guide photocopiable resources
Use pages 45–47 to consolidate key concepts.

Workbook
See pages 32–37 for additional supportive activities.

Unit-by-unit notes

Unit 7: England

In this unit, pupils learn:
- about the physical and human geography of England
- how photographs, maps and words can give you information about a place
- how to investigate the quality of life.

England is the largest country in the United Kingdom. It is about the same size as Scotland, Wales and Northern Ireland put together, but has four times the population. London is the biggest city, with nearly ten million inhabitants. It is a worldwide centre for banking and commerce, and Heathrow airport is the busiest airport in Europe. Despite the changes of the last few decades, England still has a significant industrial base. Engineering, chemicals, textiles, aircraft, cars and electronics are some of the main industries. There are considerable regional differences, particularly between the old industrial areas of the north and the south where finance, banking and technology dominate.

Lesson 1: INTRODUCING ENGLAND

What is England like?
There is a balance of physical and human geography in this lesson. The aim is to provide an brief overview of England which pupils can build on.

Discussion
Pupils can build their understanding through talk by working closely with the maps and text.

Investigation
The images might focus on different themes such as buildings, coastlines or heritage sites.

Mapwork
Many English counties have a long history. They are ancient geographical areas and have served as centres of administration over many centuries. Exploring historical links could add an extra dimension to this activity. Always follow the latest advice for practising online safety in research activities.

Lesson 2: FINDING OUT ABOUT SANDWICH

How has Sandwich developed?
Sandwich is an ancient town which has preserved many medieval and historic features. Once an important port, it fell into decline as the river which linked it to the sea silted up. Refugees from the Low Countries settled in Sandwich in large numbers in the 17th century. In the 18th century, the first market gardens in England were set up in the area following the Dutch example.

Discussion
You might want to prompt pupils to think about physical changes as the town stopped being a port, historical migration, population and the changing economy.

Mapwork
A guided walk or trail is a way to explore a locality, and it will provide a context for pupils to develop their mapwork and other related geography skills.

Investigation
If you find it difficult to access information relating to your immediate area, pupils could devise a timeline for their region and focus on a number of different places.

Lesson 3: LIVING IN SANDWICH

How has Sandwich changed?
The science park which has developed on the edge of the town is a major new development and includes a range of hi-tech industries. The vertical farm and renewable energy plant are two other ventures which aim to be environmentally friendly. How this area has developed is another example of new uses for old sites – a theme explored in Unit 4.

Quality of life

Discussion
People have different views about the changes happening in and around Sandwich. These are explored in the speech bubble comments. Ask pupils why they think these new developments were made, and ask them to give reasons for their views.

Investigation
This activity encourages pupils to think about local changes and possible future developments. Change can be a difficult concept for children who tend to think that places have always been the same.

Teacher's Guide photocopiable resources
Use pages 48–50 to consolidate key concepts.

Workbook
See pages 38–43 for additional supportive activities.

Unit-by-unit notes

Unit 8: Europe

> **In this unit, pupils learn:**
> - about the different countries and landscapes of Europe
> - why the European Union was formed
> - what makes Europe special.

Europe is one of the smallest but also one of the most varied continents. It stretches from Portugal and Ireland in the west to the Ural Mountains in the east. The coastline is deeply indented and rich in natural harbours. Seas, such as the Mediterranean, Adriatic and Aegean, have served as the focus for many ancient civilisations. From the 16th to the 20th centuries, Europe spearheaded the Industrial Revolution. Today, although eclipsed by the US and China, Europe still retains a key position in terms of industry, agriculture, trade and finance. Europe is also helping to forge new approaches to environmental thinking as people grapple with the problems of living within the finite carrying capacity of the planet.

Lesson 1: INTRODUCING EUROPE

What are the regions of Europe?
This lesson uses a map to introduce the physical landscape of Europe. The portraits of the three children illustrate regional differences.

Discussion
There are many differences to pick up on, including climate and landscape, from the cold of the northern tundra to the heat of Mediterranean lands.

Mapwork
Italy is one example, with mountains, forests, and Mediterranean landscapes.

Investigation
Pupils might identify a number of different routes between the two cities. You might extend the exercise by tracing routes between other cities such as London and Istanbul, which, whilst they are not marked on the map, are easy to locate.

Lesson 2: THE EUROPEAN UNION

How can countries work together?
The European Union is an interesting example of international co-operation. Some people have doubts about the way it works; others agree that lots of social, environmental and economic problems require international rather than national solutions.

Discussion
You might want to pick up on the idea that there can be a tension between co-operation and a country wishing to create its own rules around things like trade and standards.

Mapwork
Once pupils have made their list from the map on Pupil Book page 46, they could check their answers by looking at a list online. Always follow the latest advice for practising online safety in research activities.

Investigation
You could extend this exercise by thinking of reasons for not joining the EU as a prelude to a class debate on the pros and cons.

Lesson 3: CELEBRATING EUROPE

What is special about Europe?
You could use the information on Pupil Book pages 48 and 49 as the basis for your own class celebration for your continent. In doing this you can decide whether you want to keep a focused geography focus or whether to adopt an interdisciplinary approach.

Mapwork
It will help the children if their maps are in colour. You may have some old maps that are no longer needed which pupils can use. Colour photocopies or online colour maps offer viable alternatives.

Investigation
Talk with the children about the different criteria they could use for selecting their 'wonders'. Plants, animals, buildings, museums and leisure facilities are some of the possibilities.

> **Teacher's Guide photocopiable resources**
> Use pages 51–53 to consolidate key concepts.
>
> **Workbook**
> See pages 44–49 for additional supportive activities.

Unit-by-unit notes

Unit 9: South America

In this unit, pupils learn:
- why the Amazon rainforest is so special
- how it is threatened
- how it can be protected.

Rainforests extend approximately ten degrees north and south of the equator. Although they only cover a relatively small proportion of the Earth's surface, they contain well over half of all plant and animal species. Most of these have never been studied or named by scientists. Amazonia is the most important and largest remaining rainforest region in the world. It is the ancestral home to around one million indigenous people. Although many have moved to cities, there are still thought to be some groups that have never been in contact with the outside world.

Lesson 1: INTRODUCING THE AMAZON

What is the Amazon like?
Many pupils will have heard of the Amazon, but they may not be aware of where it is in the world or even that it is on the equator. Emphasise its size: Amazonia covers an area more than 20 times the size of the UK.

Why is the rainforest being cleared?
For countries like Brazil, Amazonia is a frontier zone waiting to be exploited and developed. It is rich in mineral resources, has huge potential for hydroelectric power and can provide space for landless workers. As the forest is cleared, the land is being used for cattle ranching and growing soy beans. The wood from the forests also earns valuable export income.

Mapwork
This activity sets the Amazon in an international context.

Investigation
Pupils could include charts, diagrams and other visual information in their fact files.

Lesson 2: USING THE RAINFOREST

What is it like to live in the rainforest?
The example of rainforest living on Pupil Book page 52 outlines a traditional way of living which uses the forest in a sustainable manner. This contrasts with the destructive uses described in the previous lesson.

Discussion
Once pupils have discussed the questions, see if you can draw out an understanding of sustainable ways of working with natural resources as compared to the destructive methods outlined in the previous lesson.

Why is the rainforest so important?
Questions about the pros and cons of clearing the rainforest lie behind the information on this page and might lead to an in-depth class project if there is time.

Investigation
This activity considers the immediate and local effects of forest clearance. However, the rainforest also helps to regulate world climate. If more than half of it is cleared, it is believed that this could trigger changes in the climate that will affect both remaining forest areas and planetary systems.

Climate change
Explain to the children how the Amazon and other tropical rainforests are like the Earth's lungs. As well as helping to keep the climate stable, the trees also store huge quantities of carbon.

Lesson 3: SAVING THE AMAZON

What was Chico Mendes trying to do?
Chico Mendes (1944–88) is an iconic figure in rainforest conservation. A rubber tapper who worked as a labourer in the Amazon rainforest, he successfully called for the government to set up reserves in which only sustainable harvesting would be permitted. The idea of attributing economic value to conservation areas broke new ground at the time. Chico Mendes also realised the importance of championing the rights of local and indigenous people – the natural guardians of the forest. His campaigns were strongly resisted by landowners, but he continued to fight for the rainforest until his death. Please note the upsetting circumstances around Chico Mendes' death if you wish for pupils to carry out further research.

Discussion
This is a good opportunity to talk about the importance of role models and influencers. Although children may feel powerless in the face of environmental problems, it is important to remember that we can all make a difference. The campaigner Greta Thunberg is living proof of just what can be done.

Mapwork
Pupils will need blank world maps to complete this. They may be able to distinguish between rainforest areas that have been cleared and areas that remain intact if this information is readily available. Always follow the latest advice for practising online safety.

Investigation
Pupils could illustrate each panel of their timelines.

Teacher's Guide photocopiable resources
Use pages 54–56 to consolidate key concepts.

Workbook
See pages 50–55 for additional supportive activities.

Unit-by-unit notes

Unit 10: Asia

In this unit, pupils learn:
- how Southeast Asia is changing
- about different aspects of Singapore
- how Singapore is planning for the future.

Southeast Asia is a geographical region lying wholly within the tropics. With the mainland of Asia to the northwest and the Pacific Ocean to the east, Southeast Asia occupies a strategically important location. Air and shipping routes are focused on Singapore and the Malay peninsula. The coastal areas are studded with islands. Indonesia and the Philippines are both on tectonic plate boundaries and are the scene of considerable volcanic activity. Since 1967, the countries of Southeast Asia have been linked together through ASEAN (the Association of South East Asian Nations). Rapid industrialisation and export-led growth has led to some of these countries being described as 'tiger economies'.

Lesson 1: SOUTHEAST ASIA

What is Southeast Asia like?
This lesson introduces themes from both physical and human geography. A key feature is the rapid development which has occurred over the last 50 years. It is important that pupils appreciate that, although Southeast Asia was historically part of European empires, it exhibits the features of a modern industrialised world today.

Discussion
Encourage pupils to work closely with the map, pictures and text to identify landscape features and vegetation.

Mapwork
The largest cities include Singapore, Jakarta, Bangkok, Manila, Kuala Lumpur, Yangon and Ho Chi Minh City.

Investigation
Remind pupils that the fact files can include graphical as well as numerical information.

Lesson 2: INVESTIGATING SINGAPORE

What is Singapore like?

Discussion
The growth and development of Singapore in many ways symbolizes the growth and development which characterizes many parts of Southeast Asia. The swampy and unhealthy islands at the tip of the Malay peninsula have now become a thriving modern nation. Chinese, Malayan and Indian communities have all contributed to Singapore's success. Trade links with the rest of the world and the associated banking and finance industries are crucial to its continuing prosperity.

Investigation
In their advertisements pupils might want to stress the advantages of Singapore's location on the sea routes from Europe to China and Japan and its proximity to Australia. For a case study of what attracts a modern business to Singapore, find out why Rolls Royce has invested in its Seletar facility.

Mapwork
The fact that Singapore lies close to the equator is a natural opportunity to find out about other cities with a similar latitude.

Lesson 3: A SINGAPORE FAMILY

What is it like to live in Singapore?
The case study of a Singapore family traces the fortunes of one family over three generations. Like many others, they moved to escape war. Their success mirrors the growth and prosperity of Singapore. Remind children that personal case studies like this can only present one personal example of life in a country.

Discussion
You could encourage pupils to weigh up and discuss the problems of providing enough clean water and enough housing as well as the threat of extreme weather, for example.

Planning for the future
The shortage of land in Singapore means that there is a premium on using all resources as wisely as possible. New buildings and water are two key examples of how careful planning can bring long-term benefits.

Mapwork
Get the pupils to cut out cardboard to represent the blocks of flats so they can test out alternative positions as they devise their plans.

Climate change
Planning for climate change involves trying to anticipate uncertain weather events. How far into the future do you think people should be planning? Can they do this without widespread public support? Discuss this with the children.

Teacher's Guide photocopiable resources
Use pages 57–59 to consolidate key concepts.

Workbook
See pages 56–61 for additional supportive activities.

Photocopiable resource matrix

Unit	Photocopiable resource	Description
Restless Earth	1 Earthquakes and volcanoes	Children colour a cross-section diagram of the Earth and distinguish the different layers.
	2 Creating landscapes	Children complete a crossword puzzle and say how different forces shape the land.
	3 Rocks and soils in the UK	Children compare different UK landscapes using information in the Pupil Book.
Drinking water	4 Water, water everywhere	Children devise a bar chart to show how much water is consumed in different activities.
	5 Water supplies	Children colour a world map to show the places where many people do not have clean water.
	6 Conserving water	Children draw pictures and describe three different ways of saving water.
Climate change	7 Global warming	Children colour and complete a greenhouse effect diagram using a key, and then describe it.
	8 Unusual weather	Children create a 3D picture tower to show different types and effects of extreme weather.
	9 Responding to climate change	Children consider different items in a Net Zero exhibition.
Planning issues	10 Reasons for development	Children complete a table listing the advantages of different types of development.
	11 Old sites, new uses	Children devise land-use symbols and give brief reasons for using vacant sites in different ways.
	12 Planning game	Children compile an estate agent's brochure for their school site.
Transport	13 Travelling further, travelling faster	A survey in which children compare the advantages and disadvantages of air travel.
	14 Transport problems	A 'snakes and ladders' game to show how road improvement schemes solve some problems but create other issues.
	15 Hidden costs	A survey sheet that children can use to find out about local traffic problems.

Photocopiable resource matrix

Aim	Teaching points
To consolidate pupils' understanding of what lies beneath the Earth's surface.	Check that pupils understand the scale of the diagram – it is 6000 km to the centre of the Earth.
To reinforce understanding of different processes of erosion.	Children will need to read the text in the Pupil Book carefully before writing their ideas.
To show that the landscapes of the UK were formed in different ways.	Extend the work by finding the different landscapes in other parts of the UK, or in your region.
To illustrate the range of ways that we use water in our daily lives.	Activities such as brushing teeth and drinking will show up on the graph as a vertical line as they won't fill a whole block.
To highlight how access to clean water is a worldwide problem.	Discuss the problems caused by polluted water and possible solutions.
To draw attention to ways of helping to solve water-supply problems.	Make sure that the children understand why different solutions are needed in different places.
To illustrate and reinforce understanding of the greenhouse effect on the world's climate.	Children can refer to the diagram and supporting description in the Pupil Book for support.
To illustrate the impacts of global warming on the weather.	Provide scissors and glue for this activity. You could discuss the different types of unusual weather before children begin.
To alert pupils to what would no longer be needed if we succeed in reaching net zero.	This activity could be completed as a group or class exercise to allow pupils to pool their ideas.
To emphasise how empty sites can be developed in a variety of ways.	Ensure pupils understand how different categories overlap. For example, farming could be good for the environment, and housing developments have implications for transport.
To show that planning often involves balancing different interests and considerations.	Discuss how each site might be used before the children make their choices.
To help children see the limitations and potential of redevelopment.	It will be useful for the children to have access to a plan of the school, if possible.
To alert pupils to the way that all innovation has a hidden cost.	It will be interesting to see the results of the survey and perhaps extend the work as a class debate.
To show that traffic problems cannot be solved simply by road improvement.	If children land on the bottom of a ladder, they go up it. If they land on the tail of a snake they move back down it. Ensure that pupils understand how the snakes and ladders are linked to real-world issues.
To promote local fieldwork and investigations.	You will need to select a suitable place to conduct the survey. The children should work in groups.

Photocopiable resource matrix

Unit	Photocopiable resource	Description
Conservation	16 Threatened wildlife	Children make notes and drawings to complete a chart of endangered wildlife.
	17 Antarctica	An exercise in which children speculate on changes in Antarctica over the past two centuries.
	18 Conservation projects	Children complete a flow chart and write a short summary of what it portrays.
England	19 Introducing England	Working from the maps in the Pupil Book, children collect information about specified grid squares.
	20 Finding out about Sandwich	Children draw pictures linked to a map of some of the key features of Sandwich, a town in England.
	21 Living in Sandwich	Children select terms from a word bank to describe six different scenes and then write a short report.
Europe	22 Introducing Europe	Children complete a key and colour five Western European countries on an outline map.
	23 The European Union	Children complete a crossword puzzle to identify the countries that first formed the EU, and make a fact file about the EU.
	24 Celebrating Europe	A game in which children collect four cards relating to one European country.
South America	25 Introducing the Amazon	Using an outline world map, children identify rainforest, desert and polar regions.
	26 Using the rainforest	Children colour an outline rainforest drawing and consider the impact of forest clearance.
	27 Saving the Amazon	Children devise a speech to represent the views of either Chico Mendes or Da Silva.
Asia	28 Southeast Asia	A mapwork game in which children identify different countries in Southeast Asia.
	29 Investigating Singapore	Children compile a set of data about Singapore using different given headings.
	30 A Singapore family	Children construct a simple model illustrating futures thinking.

Photocopiable resource matrix

Aim	Teaching points
To identify species that are threatened or in serious decline.	Pupils will need to research their own examples in order to complete the activity sheet.
To raise awareness of the unique but fragile Antarctic environment.	Before pupils start this exercise, discuss what changes might have happened.
To illustrate how saving wildlife benefits both people and the environment.	Ask the pupils to think about other situations where a flow chart might be useful.
To consolidate understanding of four-figure grid references using a UK map.	Remind the children to start at the bottom left-hand corner going 'along the corridor and up the stairs'.
To show how pictures, words and maps can be combined to create a place portrait.	The children could devise a similar picture map of their own local area.
To show the features that contribute to the quality of an environment.	Pupils could use their own words to supplement the terms in the word bank.
To gain familiarity with the political map of Western Europe.	Suggest that pupils could select a single colour for those countries not named on the map.
To explore the countries and effects of the European Union.	Children can refer to the map in the Pupil Book for support.
To consolidate geographical knowledge in a fun game format.	Invite pupils to devise their own European countries game as an extension activity.
To contextualise the Amazon rainforest alongside other key biomes.	Invite children to name the different regions shown on the map.
To highlight the value of the rainforest environment.	There are a number of creatures 'hidden' in the drawing. Further work into rainforest animals would be a natural extension activity.
To show that there are two sides to the debate about the future of the rainforest.	You could use this activity as preparation for a role play or class assembly about the rainforest.
To consolidate children's understanding of Southeast Asia.	Note that Singapore occupies such a small territory that it cannot be represented on the map.
To build a portrait of Singapore that draws on physical, human and environmental geography.	Talk with the children about the range of information they might include under each heading.
To explore some of the ways in which Singapore is thinking ahead and building resilience.	The models could form part of a larger display on sustainability and futures thinking.

 # Earthquakes and volcanoes *Name* ....

1. Colour the empty boxes in the key.
2. Use the key to colour the diagram.
3. Write a sentence about each word.

Key	
sea	blue
land	green
crust	brown
mantle	orange
core	yellow

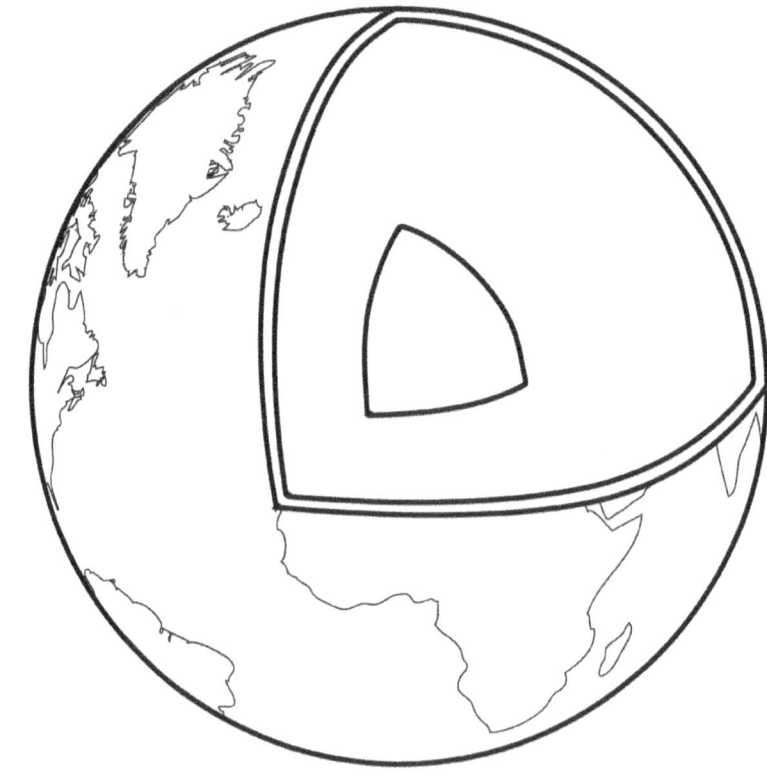

The Earth's crust _____

_____

_____

The mantle _____

_____

_____

The core _____

_____

_____

2 Creating landscapes

Name ....

1. Use the clues to help you complete the crossword.

Across

1. A current of air
2. Large streams
3. A ridge of water found in seas and oceans

Down

1. A force that breaks rock apart
2. Frozen water

2. How does each force shape the landscape?

Force	Effect on landscape
Frost	
Rivers	
Waves	
Wind	
Ice	

3 Rocks and soils in the UK   Name ....

1. Colour the drawings of the different landscapes from around the United Kingdom.

2. Explain how each landscape was formed.

Landscape	How was it formed?
Eryri (Snowdonia)	
Herefordshire	
North Sea	
Oxfordshire	

4. Water, water everywhere

Name ....

1. Add up the numbers next to the water drops to find out how much water one person might use in a day.

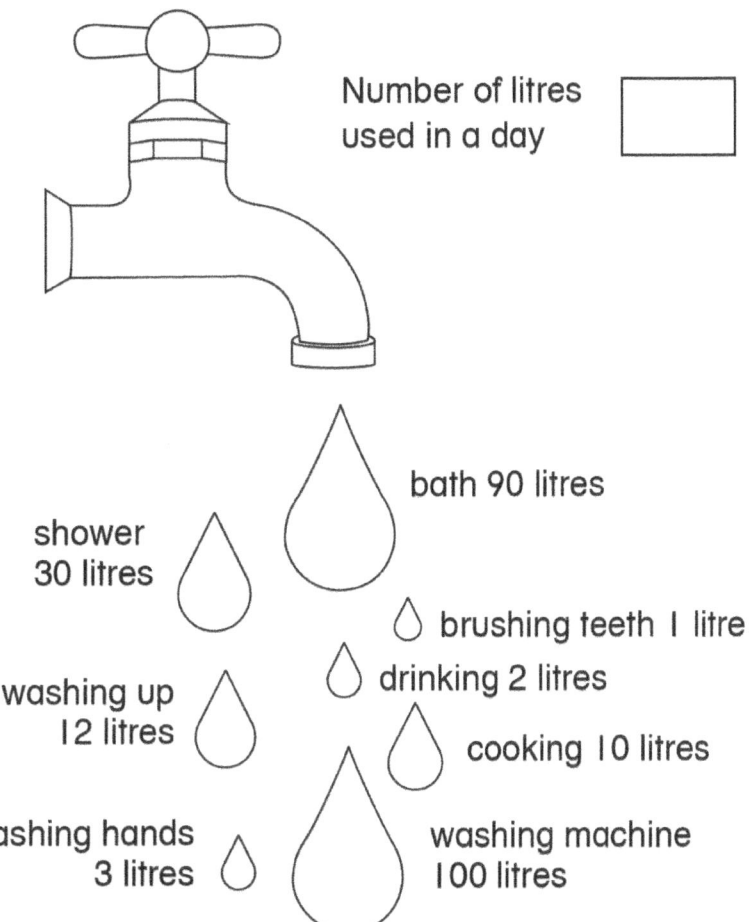

Number of litres used in a day ☐

bath 90 litres
shower 30 litres
brushing teeth 1 litre
drinking 2 litres
washing up 12 litres
cooking 10 litres
washing hands 3 litres
washing machine 100 litres

2. Make a bar chart of the amount of water used for each activity.

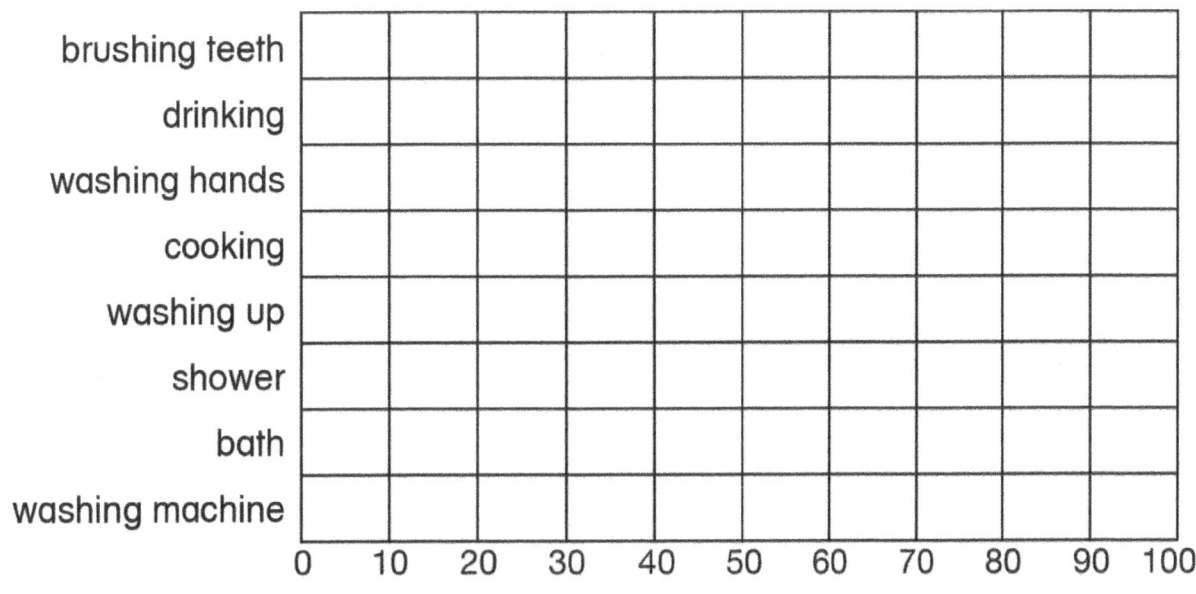

5 Water supplies

Name ....

1. Colour the world map to show the places where many people do not have clean water. Colour the key.

2. Write some of the problems caused by polluted water.

Key: Places where many people do not have clean water ☐   Other places ☐

Problems caused by polluted water
_____
_____
_____
_____
_____

6 Conserving water   Name ....

1. Draw pictures of three ways of saving water.
2. Write a few words saying how each one saves water.

	How does it save water?
	_____ _____ _____ _____ _____
	How does it save water? _____ _____ _____ _____ _____
	How does it save water? _____ _____ _____ _____ _____

7 Global warming

Name ....

1. Colour the empty boxes in the key.
2. Colour and complete the greenhouse effect diagram.
3. Now write a few sentences explaining what the diagram shows.

Key	
Heat from the sun	
Heat from the earth	
Greenhouse gases	

_____
_____
_____
_____

8 Unusual weather

Name ....

1. Draw a picture in each of the panels and write what it shows.
2. Cut around the edges, fold along the dotted lines and glue your 'picture tower' together.

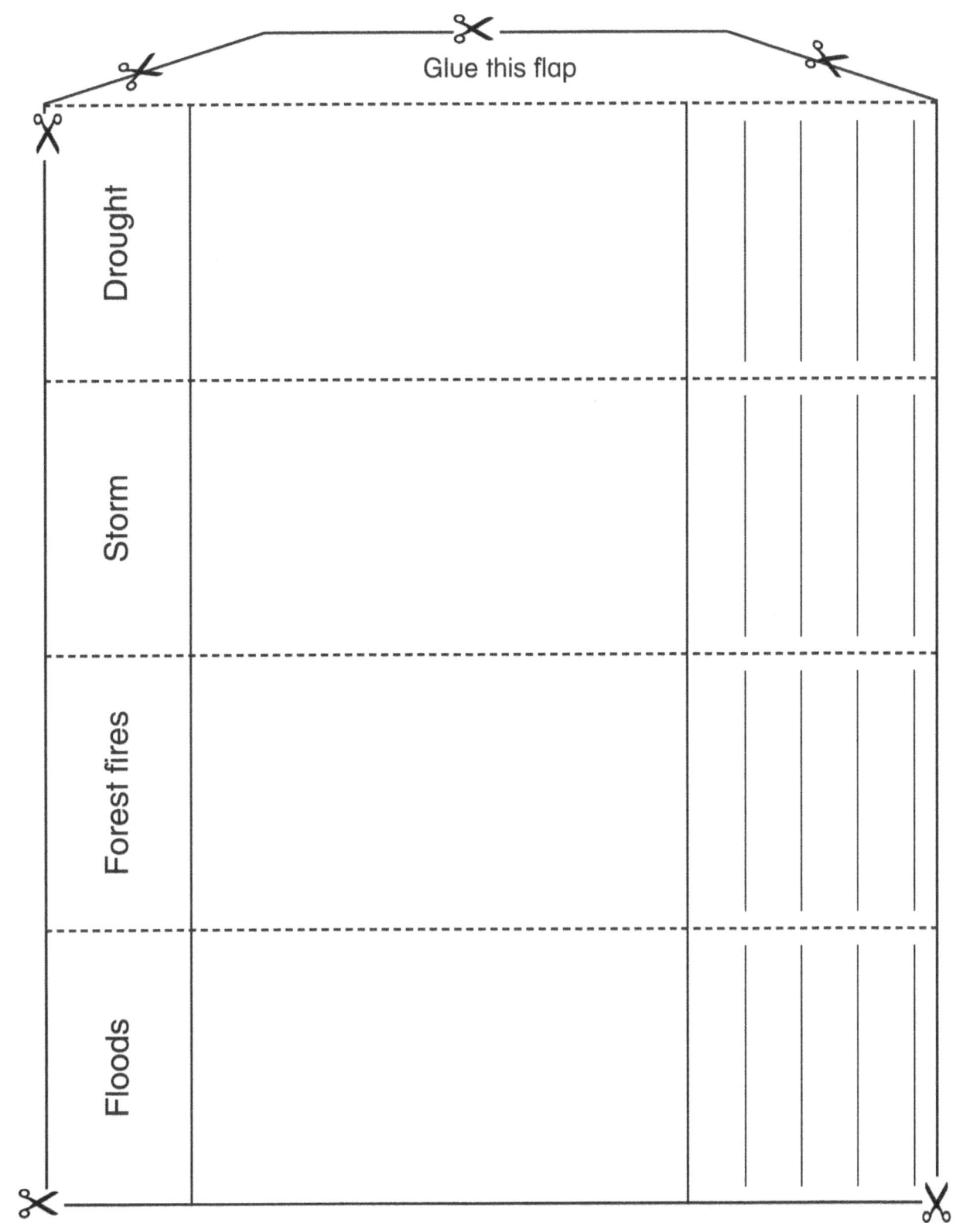

Responding to climate change Name ....

1. Say why you think these items were put in the Net Zero exhibition.

Item	Why it was put in the exhibition
Coal	
Plastic toys	
Throwaway clothes	
Bottled water	
Petrol engines	
Disposable nappies	

2. Draw two net zero icons in the empty boxes.

10 Reasons for development

Name ....

Look at the drawing. Write a sentence about the advantage of each possible use.

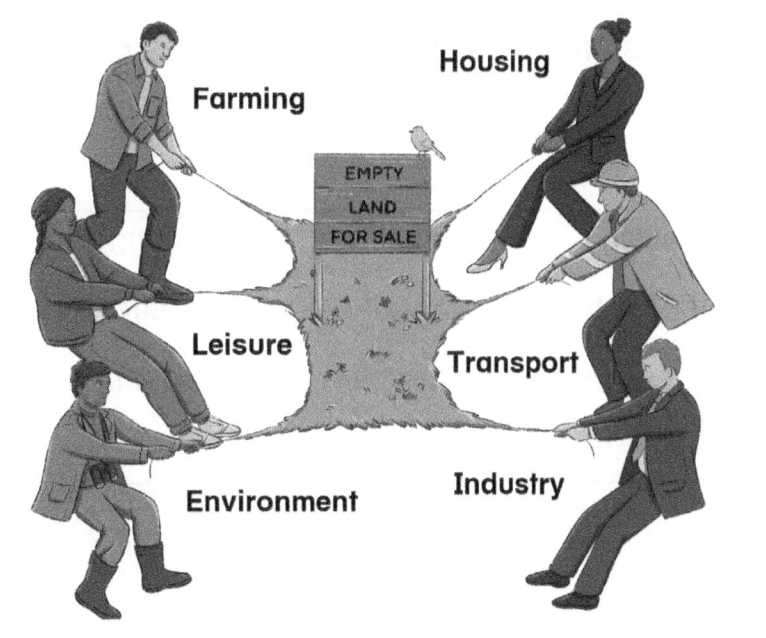

Possible use	Advantage
Farming	
Housing	
Leisure	
Transport	
Environment	
Industry	

Old sites, new uses

1. Draw symbols to show different ways of using land.

housing	industry	farming	leisure	environment

2. What is the best way of using each site on the plan below?

3. Draw symbols on the plan and write a sentence in the table giving your reason.

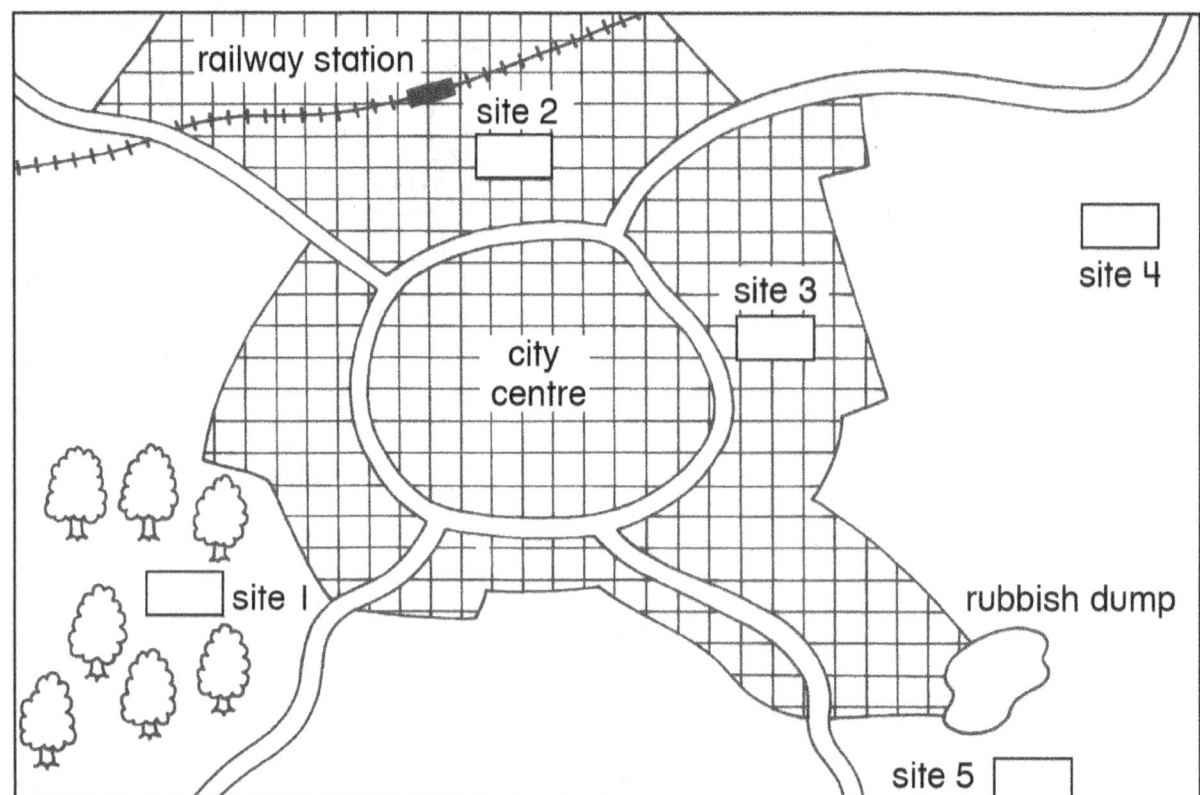

Site	Reasons for your choice
Site 1	
Site 2	
Site 3	
Site 4	
Site 5	
Site 6	

12 Planning game

Name ....

Write the information you would put in a sale brochure for your school.

Name of school
_____

Number of rooms _____

Open spaces/grounds _____

Any special facilities _____
_____

Possible uses _____
_____

How can the site be reached by car or public transport?

_____
_____
_____
_____
_____
_____
_____

Plan of the school

13 Travelling further, travelling faster Name ....

1. Complete the drawings and descriptions of air travel in the table below.
2. Circle one of the scores in each row.

Advantages	Description	Plus score
	Flying is the safest way to travel.	OK +1 good +2 excellent +3
	Flying is very fast.	OK +1 good +2 excellent +3
		OK +1 good +2 excellent +3
Disadvantages	**Description**	**Minus score**
	Not good for heavy and bulky goods.	annoying −1 bad −2 serious −3
		annoying −1 bad −2 serious −3
	Planes are very noisy.	annoying −1 bad −2 serious −3

3. Add up your plus and minus scores. + ☐   − ☐
4. Which total is larger? What does this show you about your view of flying?

_____

_____

14 Transport problems

Name ....

1. Colour the snakes and ladders game below.
2. Play the game with a partner. You will need a 1–6 spinner and a counter for each player.

FINISH	55	Better roads create more traffic.	53	52	Shortage of oil and petrol.	50
43	44	45	Traffic noise causes stress.	47	48	49
42	41	40	39	38	37	Park and ride scheme.
29	30	31	32	33	34	Fumes cause health problems.
Bypss built around villages.	27	26	25	24	23	22
15	16	17	Motorways link large cities.	19	20	21
14	13	12	11	Increase in road accidents.	9	8
START	New traffic schemes improve traffic flow.	3	Ring road around city centres.	5	6	7

15 Hidden costs

Name ....

Make a survey of the traffic problems in your area.

1. Ask passers-by which traffic problem they think is most serious.
2. Colour a square to show their answer.
3. Write a few sentences about what you have discovered from the survey.

Traffic Problem	1	2	3	4	5	6	7	8	9	10	11	12
Noise from traffic												
Not enough crossing places												
Traffic travelling too fast												
Roads without pavements												
Not enough safety barriers												
Too many parked cars												
Shortage of cycle routes												
Exhaust fumes												
Heavy lorries												
Traffic jams												

_____

_____

_____

_____

16 Threatened wildlife

Name ....

1. Colour the drawings of threatened creatures.
2. Write why each one is threatened.
3. Add three more examples of your own.

Whales	Eagles	Butterflies
Orchids	Rhinos	Teak trees

17 Antarctica

Name ....

1. Draw what the penguins might have seen 200 years ago.

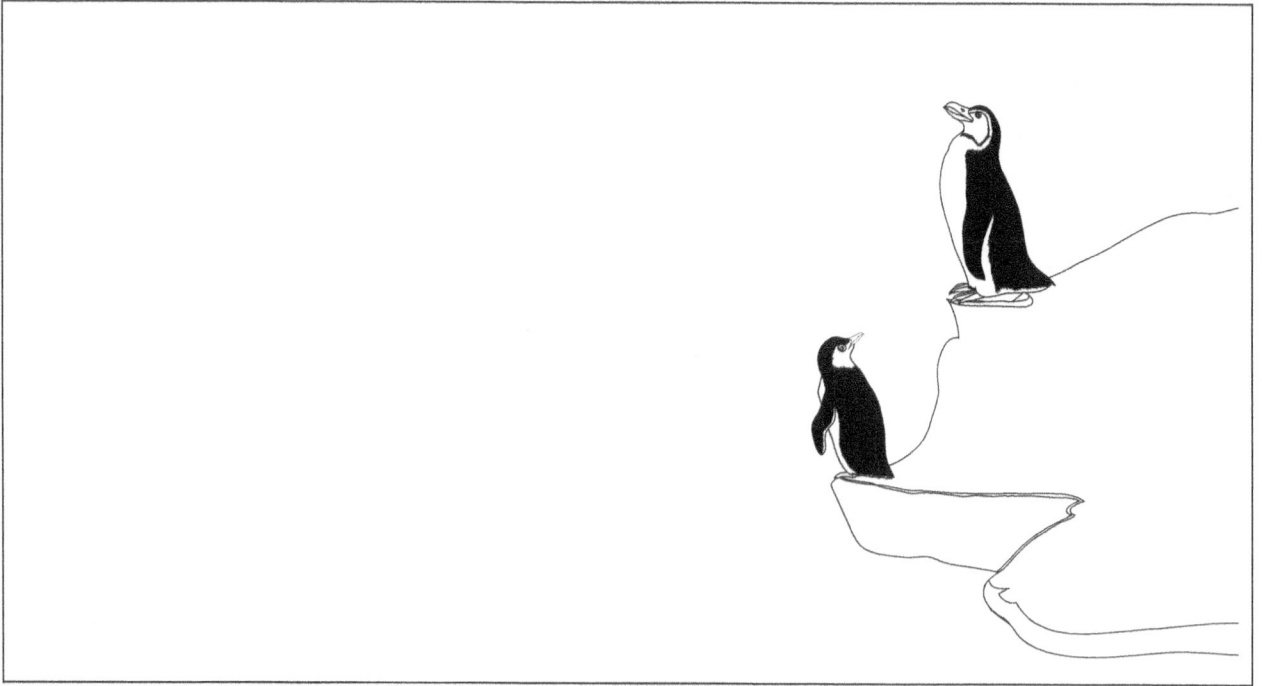

2. Draw what they might be looking at today.

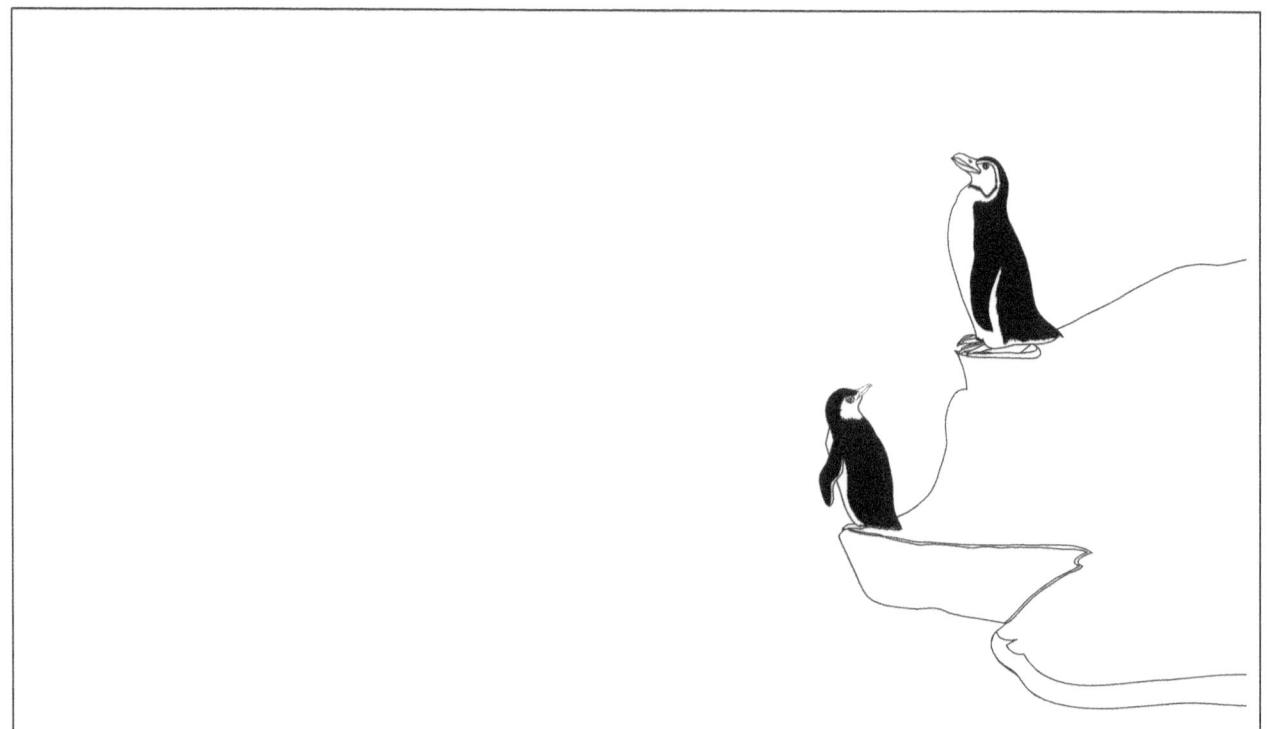

3. Label your drawings using green for natural features and red for things brought in by people.

18 Conservation projects

Name ....

1. Draw pictures in the flow chart to show how monarch butterflies have been saved from danger.

Millions of butterflies fly north in the spring.	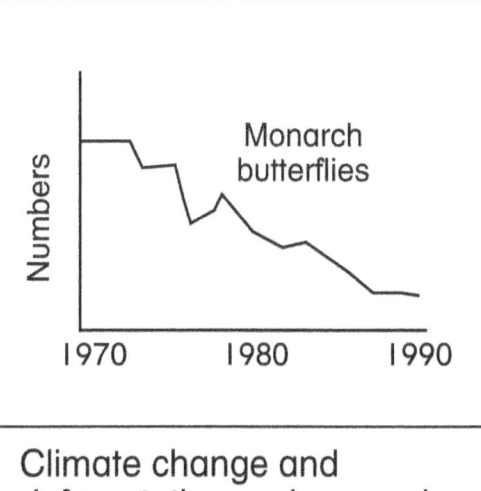 Climate change and deforestation endangered the butterflies.
Today eco-tourists come to see these amazing butterflies.	Local people have new jobs and have stopped cutting trees.

2. How have local people, trees and the butterflies all benefited from eco-tourism?

_____

_____

19 Introducing England

Name ....

1. Look at each grid square. They are numbered from the bottom left-hand corner.

2. Use the maps on Pupil Book pages 38–39 to find one feature in each grid square.

3. Mark these features on the map.

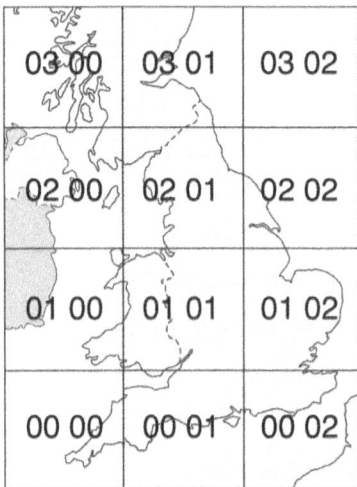

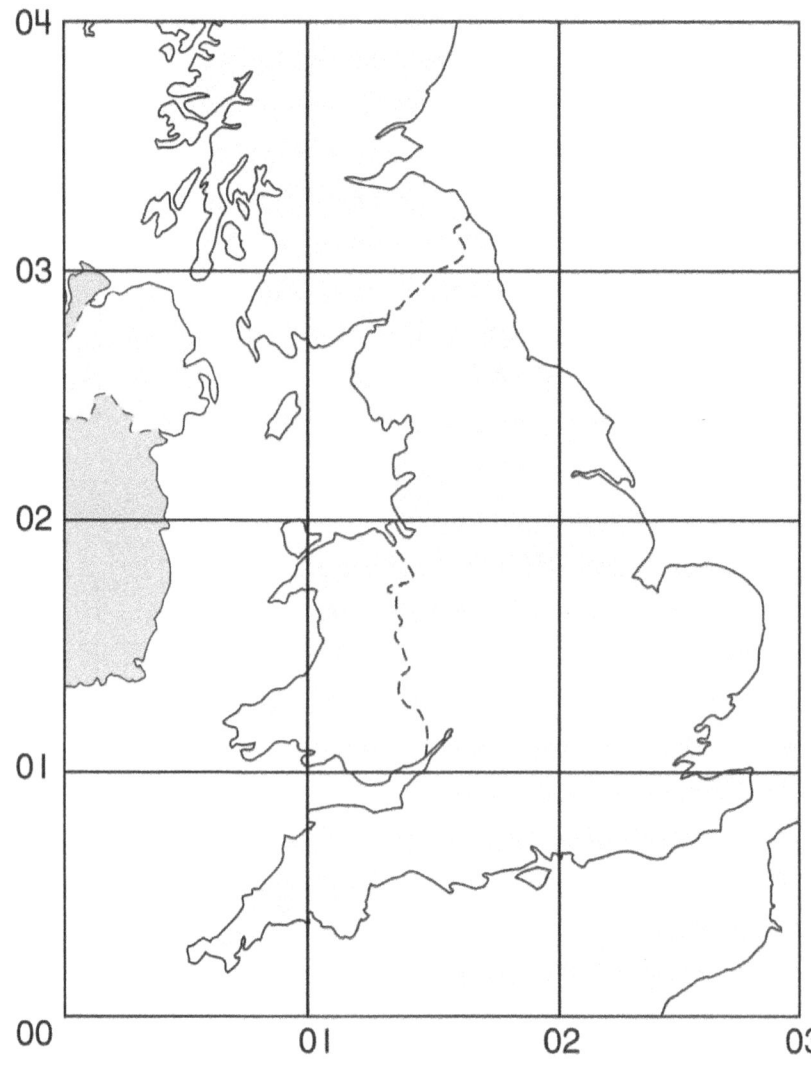

Square 00 00
Square 00 01
Square 00 02
Square 01 00
Square 01 01
Square 01 02
Square 02 01
Square 02 02

20 Finding out about Sandwich   Name ....

1. Colour the river, roads and old town centre on the map.
2. Draw pictures of the different features in the boxes.

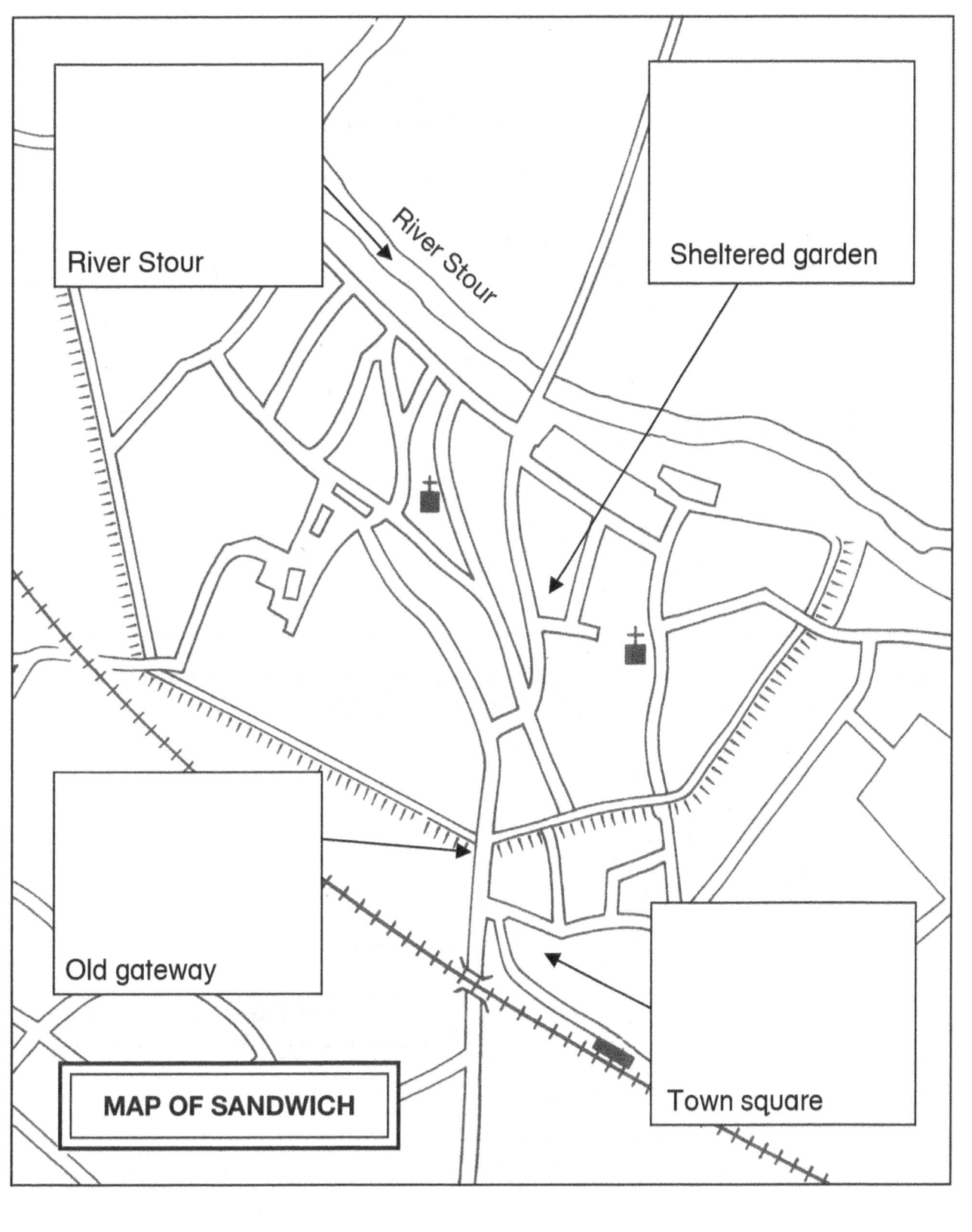

MAP OF SANDWICH

21 Living in Sandwich

Name ....

1. Write one of the words from this list under each picture.

   noisy   beautiful   dull   smoky   interesting   clean
   smelly   peaceful   unattractive   quiet   convenient

New homes | The River Stour | The Town Hall

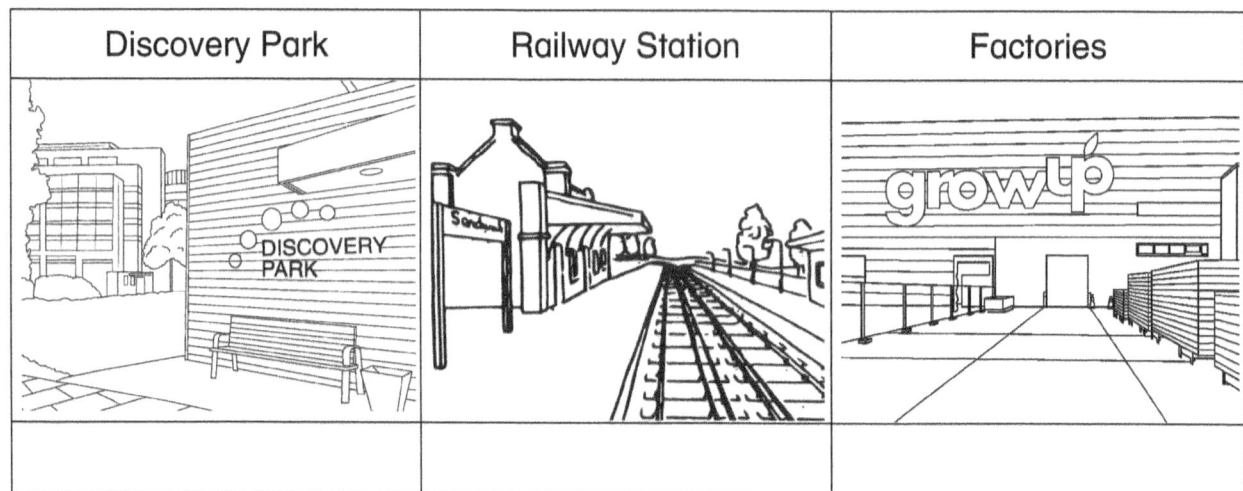

Discovery Park | Railway Station | Factories

2. Write a short report about living in Sandwich.

_____
_____
_____
_____

22 Introducing Europe

Name ....

1. Colour the code boxes in the key.
2. Colour the countries on the map.
3. Name the capital cities.

Country	Code	Capital
UK	red	
France	green	
Spain	orange	
Germany	yellow	
Italy	brown	

23 The European Union

Name ....

1. Complete the crossword of countries that first formed the European Union.

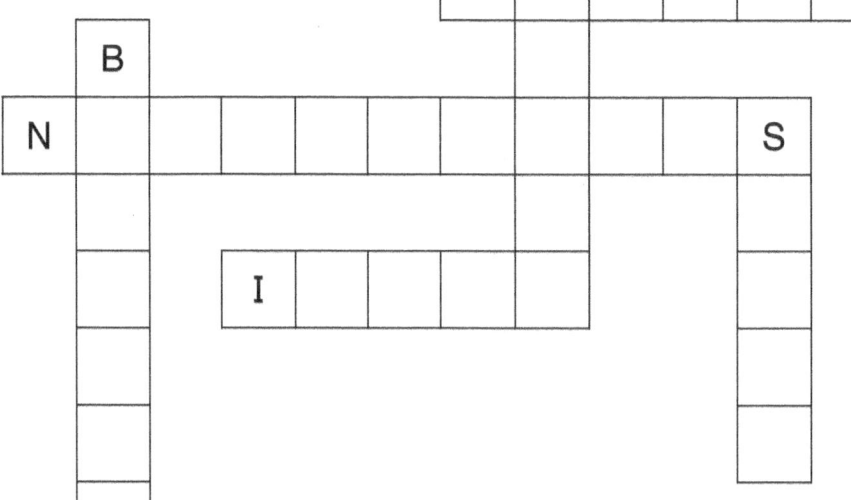

2. Make a fact file about the European Union and what it has achieved.

FACT FILE

1. _____

2. _____

3. _____

4. _____

24 Celebrating Europe

Play the countries game in groups of three.

1. Colour the cards, cut them out and put them face down on the table.
2. Each player has to say which country set they want to collect.
3. Take turns to pick up a card. Keep it if it matches your country. Put it back if it is not in your set.
4. The winner is the first to collect a complete country set.

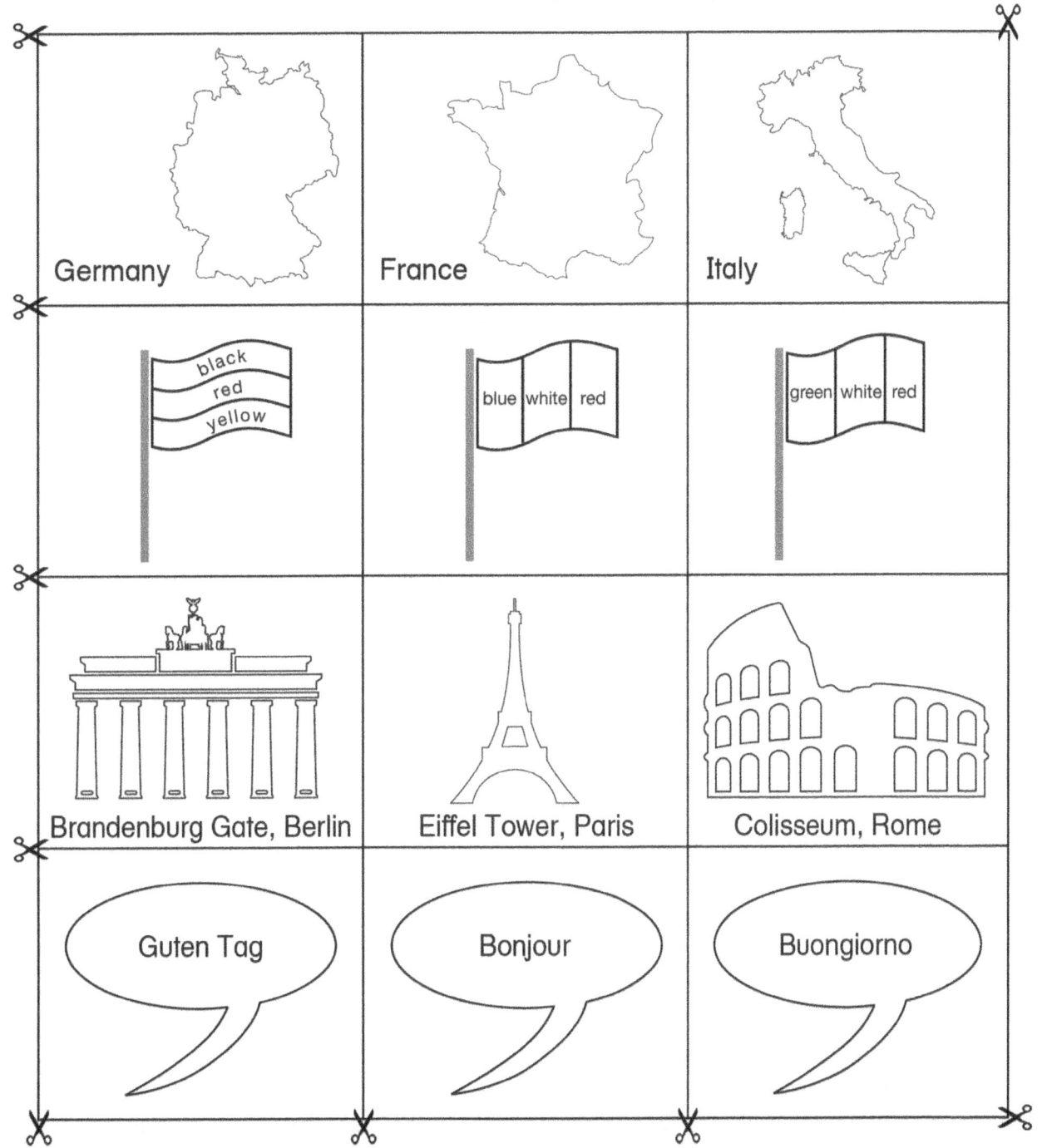

25 Introducing the Amazon

Name ....

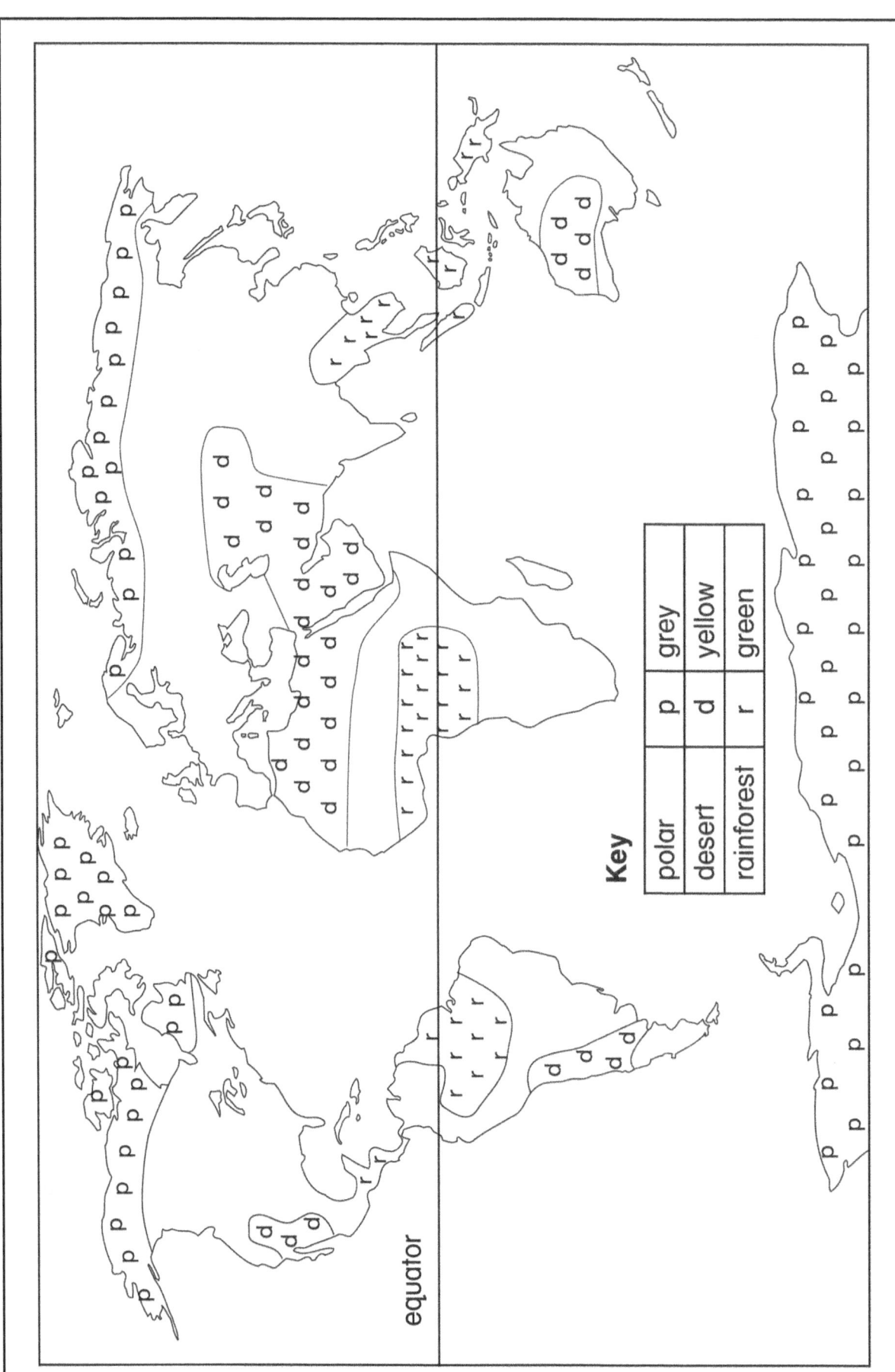

1. Colour the map to show rainforest, desert and polar regions.
2. What two things does this tell you about the Amazon (a) _____ (b) _____

26 Using the rainforest

Name ....

1. Colour the rainforest picture.
2. What would happen if the trees were cut down?
   Write a few sentences on the lines under the picture.

_____

_____

_____

27 Saving the Amazon

Name ....

Hold a debate about changes in the rainforest.

1. Ask the teacher if you are Chico Mendes or Da Silva.
2. Colour your person.
3. Make a list of the things you want to say in your speech.
4. Design a slogan for the banner.

Chico Mendes
Rubber tapper

Da Silva
Landowner

Ideas for my speech

Banner

28 Southeast Asia

Name ....

1. Play a game with a partner. Take turns to spin a 1–6 spinner. The winner is the person to land on all the countries.

Country	Tick each time you land	Key colour
1 Myanmar (Burma)		red
2 Thailand		orange
3 Vietnam		yellow
4 Cambodia		purple
5 Laos		brown
6 Miss a turn		

2. Now colour the key boxes and the countries on the map.
3. Colour all the other countries green.

29 Investigating Singapore

Name ....

Make a data and information sheet about Singapore.

Map	History

Climate	Communication

Trade and industry	Environment

30 A Singapore family

Name ....

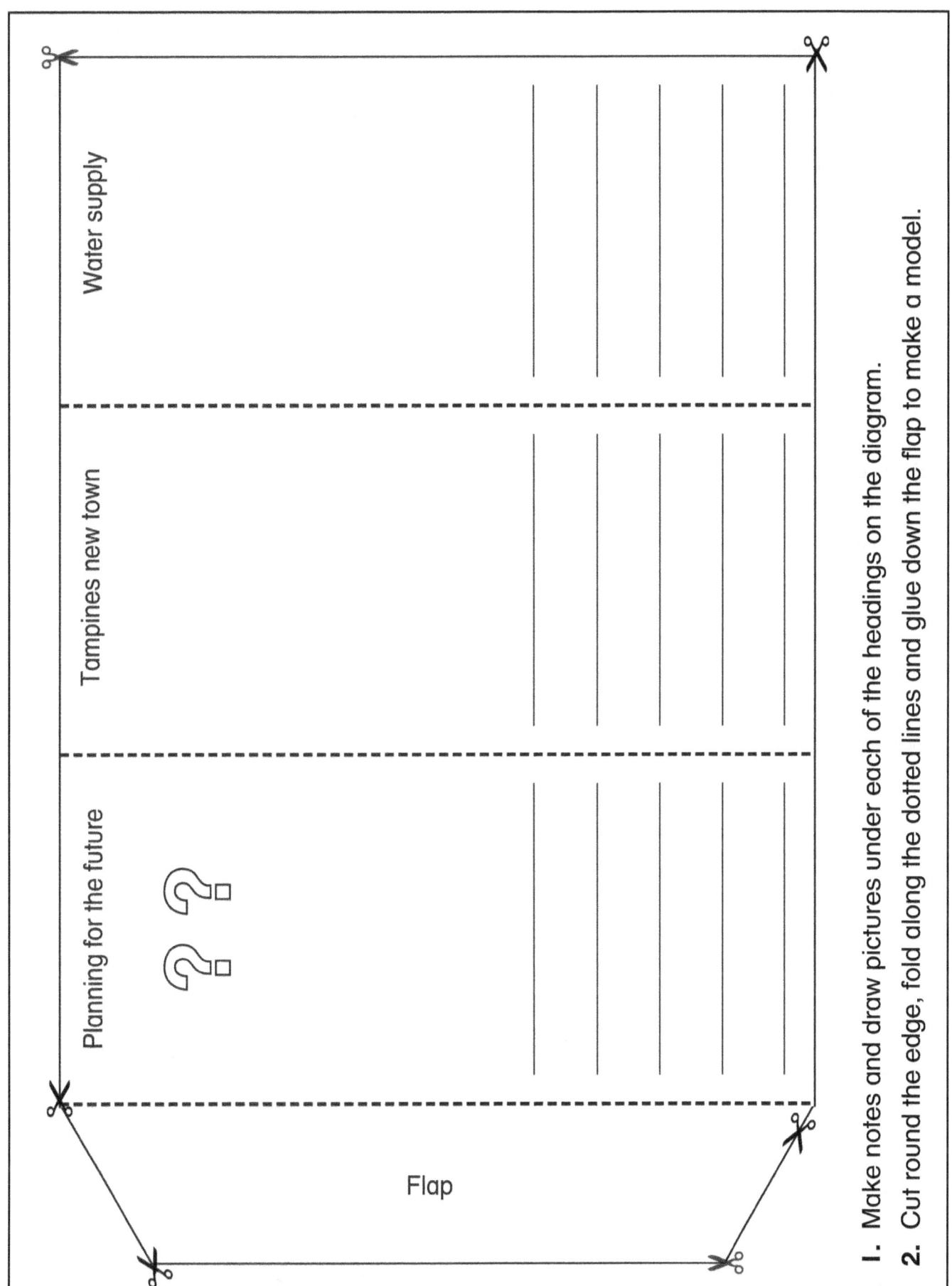

1. Make notes and draw pictures under each of the headings on the diagram.
2. Cut round the edge, fold along the dotted lines and glue down the flap to make a model.

Geography in the National Curriculum in England

The National Curriculum in England provides a geography framework for schools to follow but leaves teachers considerable scope to select and organise the content according to their individual needs. It should also be noted that the curriculum is only intended to occupy a proportion of the school day and that schools are free to devise their own studies in the time that remains.

Purpose of study

The aim of geographical education is clearly articulated in the opening section of the Programme of Study, which states:

A high-quality geography education should inspire in pupils a curiosity and fascination about the world and its people that will remain with them for the rest of their lives. Teaching should equip pupils with knowledge about diverse places, people, resources and natural and human environments, together with a deep understanding of the Earth's key physical and human processes. As pupils progress, their growing knowledge about the world should help them to deepen their understanding of the interaction between physical and human processes, and of the formation and use of landscapes and environments. Geographical knowledge, understanding and skills provide the frameworks and approaches that explain how the Earth's features at different scales are shaped and interconnected and change over time.

Subject content

The National Curriculum provides the following general guidance for each Key Stage:

Key Stage 1

Pupils should develop knowledge about the world, the United Kingdom and their locality. They should understand basic subject-specific vocabulary relating to human and physical geography and begin to use geographical skills, including first-hand observation, to enhance their locational awareness.

Key Stage 2

Pupils should extend their knowledge and understanding beyond the local area to include the United Kingdom and Europe, North and South America. This will include the location and characteristics of a range of the world's most significant human and physical features. They should develop their use of geographical knowledge, understanding and skills to enhance their locational and place knowledge.

There is an emphasis on factual and place knowledge. For example, there is a focus on learning about the UK and Europe. Map reading and communication skills are also highlighted. On the other hand, there are no specific references to the developing world, and sustainability is not mentioned directly. However, there is an expectation that schools will work from the Programmes of Study to develop a broad and balanced curriculum which meets the needs of learners in their locality. This provides schools with scope to enrich the curriculum and rectify any omissions which they may perceive.

Key Stage 2 pupils will extend their knowledge and apply their skills to areas beyond the local area, to include the UK, Europe, North and South America, Africa and Asia.

Fieldwork is covered throughout the *Collins Primary Geography* series, and consistent opportunities are provided in Investigation and Mapwork activities to 'observe, measure, record, and present the human and physical features in the local area using a range of methods, including sketch maps, plans, graphs and digital technologies', as specified in the National Curriculum.

Key Stage 2 Programme of study

Key Stage 2 Geography National Curriculum	*Collins Primary Geography* coverage
Extend knowledge of UK, Europe and North and South America	Places (all)
Location of world's most significant human and physical features	(all)
Knowledge, understanding and skills to enhance locational and place knowledge	(all)
Locational knowledge	
Locate the world's countries	(all – Mapwork)
Use maps to focus on countries, cities and regions in Europe	Places (all)
Use maps to focus on countries, cities and regions in North America	Book 4, Book 5: Places
Use maps to focus on countries, cities and regions in South America	Book 3, Book 6: Places
Name and locate counties of the UK	Places (all)
Name and locate cities of the UK	Places (all)
Geographical regions of the UK	Places (all)
Topographical features of the UK, such as hills, mountains, coasts and rivers.	Places (all)
Changing land use patterns of the UK	Places (all)
Significance of latitude and longitude	Book 6: Planet Earth
Significance of Equator, Northern and Southern Hemisphere, Tropics of Cancer/Capricorn, Arctic/Antarctic circles, Prime Meridian	Book 6: Places
Time zones	Book 4: Places
Day and night	Book 4: Places
Place knowledge	
Regional study within UK	Places (all)
Regional study in a European country	Places (all)
Regional study in North America	Book 4, Book 5: Places
Regional study in South America	Book 3, Book 6: Places
Human and physical geography	
Climate zones	Weather (all)
Biomes and vegetation belts	Environment (all)
Rivers and mountains	Book 3, Book 6: Planet Earth; Book 4, Book 5: Water; Places (all)
Volcanoes and earthquakes	Book 5, Book 6: Planet Earth; Book 5: Environment; Book 3, Book 6: Places
Water cycle	Book 5: Water
Types of settlement and land use	Settlements (all)
Economic activity including trade links	Work and travel (all)
Distribution of natural resources including energy, food, minerals, water	Book 6: Water; Work and travel (all)
Skills and fieldwork	
Use maps, atlases, globes and digital mapping	(all – Mapwork)
Use eight points of the compass	Book 5: Water
Use four and six-figure grid references	Book 3: Work and travel; Book 4: Settlements
Use symbols and keys (including OS maps)	Book 3, Book 5: Work and travel; Book 3, Book 4: Places
Fieldwork skills	(all – Investigation)

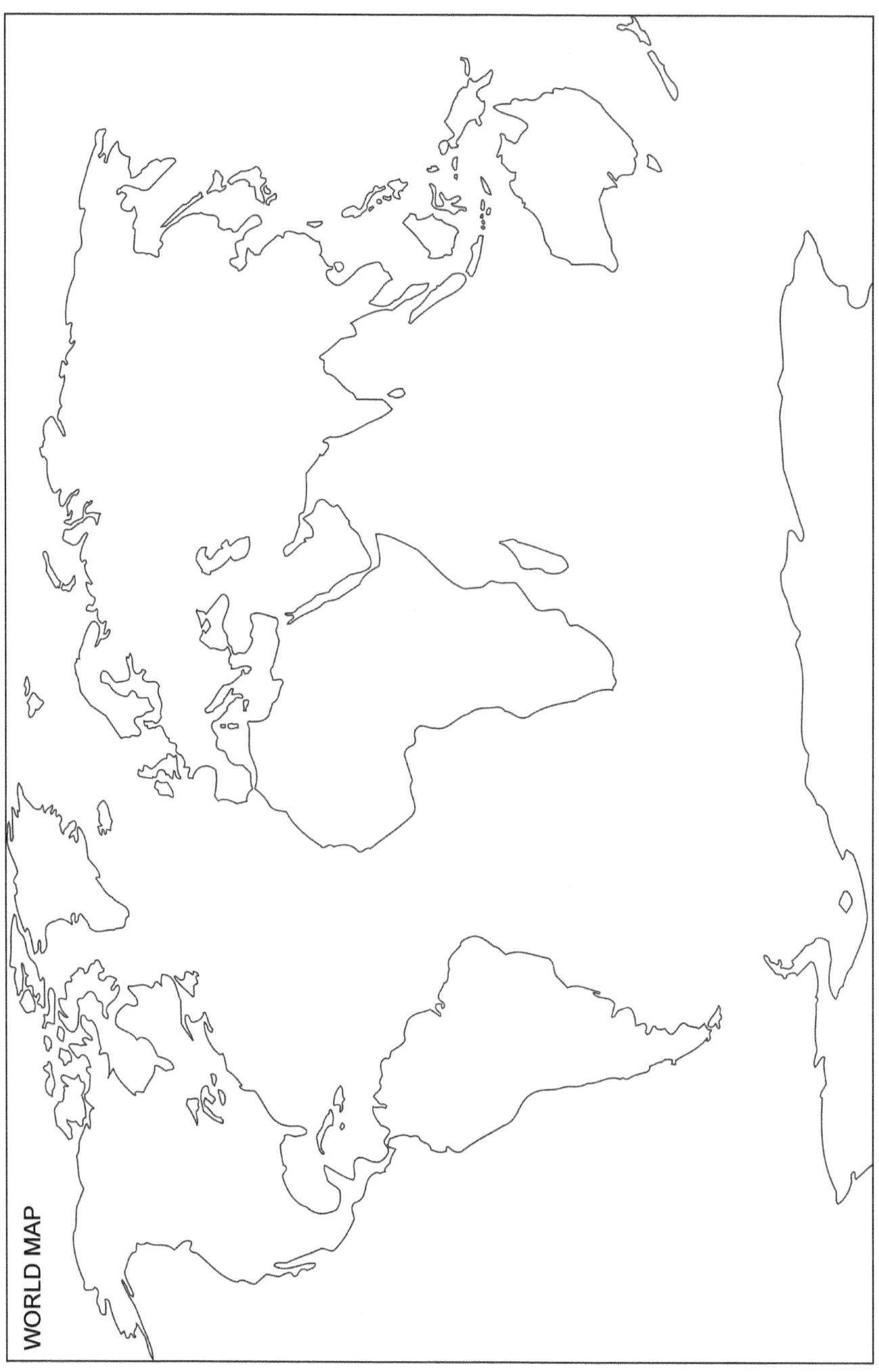

WORLD COUNTRIES

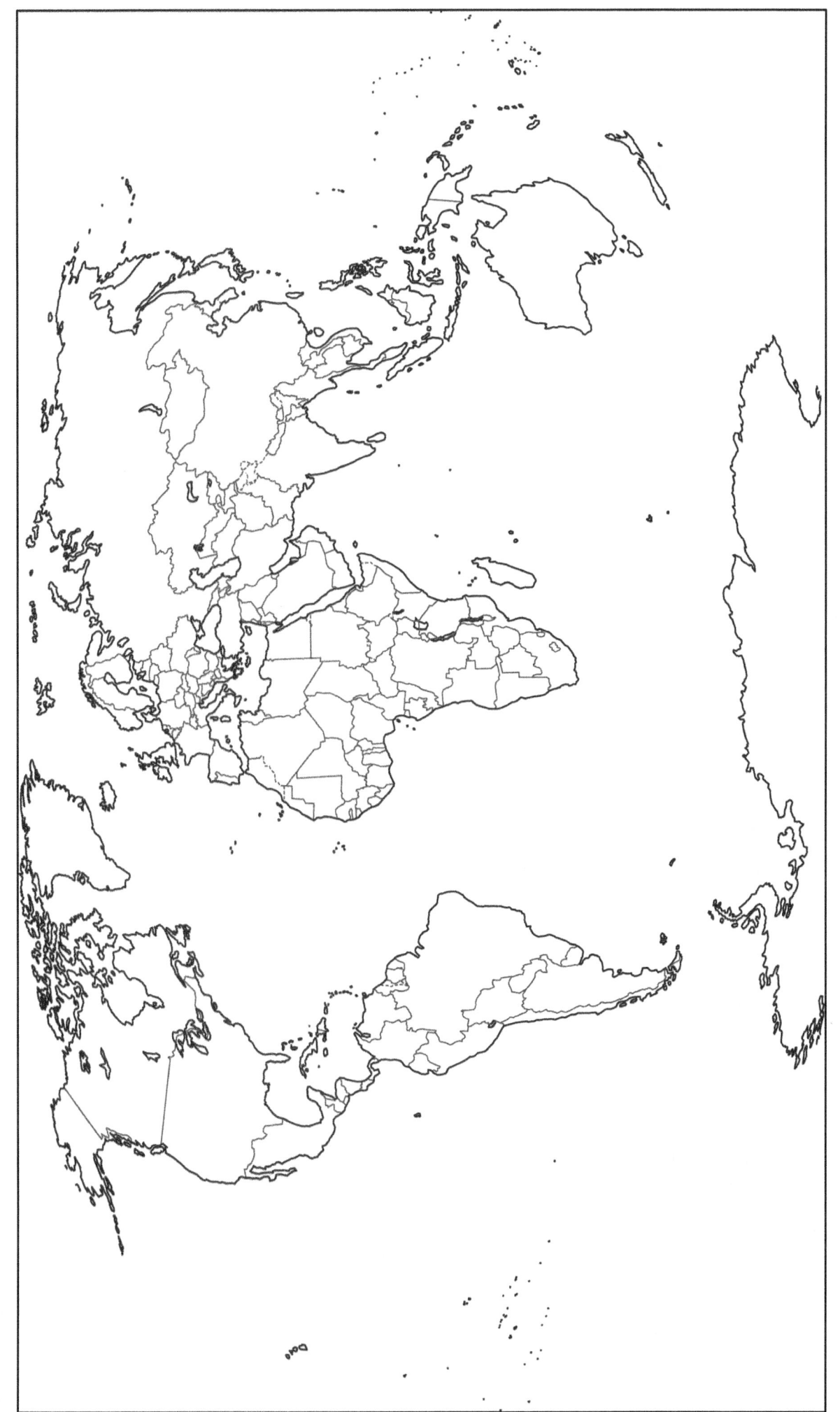

William Collins' dream of knowledge for all began with the publication of his first book in 1819.
A self-educated mill worker, he not only enriched millions of lives, but also founded a flourishing publishing house. Today, staying true to this spirit, Collins books are packed with inspiration, innovation and practical expertise. They place you at the centre of a world of possibility and give you exactly what you need to explore it.

Published by Collins
An imprint of HarperCollins*Publishers*
The News Building, 1 London Bridge Street, London, SE1 9GF, UK

HarperCollins*Publishers*
Macken House, 39/40 Mayor Street Upper, Dublin 1, D01 C9W8, Ireland

Browse the complete Collins catalogue at
collins.co.uk

© HarperCollins*Publishers* Limited 2025
Maps © Collins Bartholomew 2025

10 9 8 7 6 5 4 3 2 1
ISBN 978-0-00-872845-8

All rights reserved. No part of this publication may be reproduced, stored in a retrieval system, or transmitted in any form by any means, electronic, mechanical, photocopying, recording or otherwise, without the prior written permission of the Publisher or a licence permitting restricted copying in the United Kingdom issued by the Copyright Licensing Agency Ltd,
5th Floor, Shackleton House, 4 Battle Bridge Lane, London SE1 2HX.

British Library Cataloguing-in-Publication Data
A catalogue record for this publication is available from the British Library.

Authors: Stephen Scoffham and Colin Bridge
  (with additional original input by Terry Jewson)
Publisher: Laura White
Product manager: Natasha Paul
Development editor: Judith Walters
Proofreader: Hugh Hillyard-Parker
Cover designer and illustrator: Steve Evans
Internal illustrator: Jouve India Private Ltd
Typesetter: Hugh Hillyard-Parker
Production controllers: Alhady Ali and
  Katie Jean-Baptiste
Printed in the UK at Ashford Colour Ltd

This book contains FSC™ certified paper and other controlled sources to ensure responsible forest management.

For more information visit: www.harpercollins.co.uk/green

Acknowledgements

The publishers gratefully acknowledge the permission granted to reproduce the copyright material in this book. Every effort has been made to trace copyright holders and to obtain their permission for the use of copyright material. The publishers will gladly receive any information enabling them to rectify any error or omission at the first opportunity.

All photos: Shutterstock